Geology for Archaeologists

A short introduction

J.R.L. Allen

ARCHAEOPRESS PUBLISHING LTD
Gordon House
276 Banbury Road
Oxford OX2 7ED

www.archaeopress.com

ISBN 978 1 78491 687 9
ISBN 978 1 78491 688 6 (e-Pdf)

Front cover: Thin-section of basalt lava
Back cover: The town wall at Roman Silchester

Printed in England by Oxuniprint, Oxford

This book is available direct from Archaeopress or from our website www.archaeopress.com

To the Earth

'...no vestige of a beginning, no prospect of an end.'

James Hutton of Edinburgh (1726-1797)
Theory of the Earth

Contents

List of Figures and Tables

Author's Foreword and Acknowledgements

The purpose of this book is to introduce budding archaeologists to the elements of geology in Britain and Ireland, by means of a brief but wide-ranging text, and to encourage them to take the role of this subject in their chosen discipline more seriously than is generally the case. The reader will meet minerals and rocks, geological maps, stratigraphy and structures, and go on to consider landscapes, with an emphasis on river and coastal realms, before examining the many uses of stone and clays and the important subject of metal-mining. The later chapters finish with short case histories. These are intended to illustrate and enlarge on points introduced in the text, to reinforce the view that geology matters in archaeological work, and above all to excite and stimulate. The cases have been chosen from the British and Irish archaeological records, since it is in these islands that most field-teaching is likely to be done and professional employment to be found. Each chapter lists at the end a number of titles suitable for further reading. These lists are not intended to be comprehensive. The glossary at the end of the book elaborates on concepts and technical terms used in the text that may be unfamiliar.

It has been a challenge to write this book, but my editors at Archaeopress, Dr David Davison and Ben Heaney MA, could not have been more encouraging or helpful, with their many suggestions and careful attention to detail. It is a pleasure to thank many colleagues for their views on its structure and text, especially Professor Martin Bell and Dr Louise Jones at the University of Reading. I am indebted to Profeseor B. Burnham, Dr Mark Edmonds, Dr C. Green and Dr Ruth Shaffrey for generously providing illustrative material. I thank Michael Andrews, John Jack and Sarah Lambert-Gates at the University of Reading for their technical help with thin-sections and photomicrographs. Thanks are due to the British Geological Survey, Cornwall & Scilly Historic Environment Record, and English Heritage for permission to reproduce copyright illustrations.

J.R.L. Allen
Reading, April 2016

1. Why Geology Matters

Geology may be said to be the study of the planet Earth, the materials of which it is made, the processes acting on those materials, the products thus formed, the history of the planet since the Solar System came into being, and the life-forms that have succeeded one another on the planet since that time.

At root, humankind, one of the planet's most recent life-forms, is indissolubly a part of nature, even heaving, disconnected and disengaged city-dwellers. All people have required the bare necessities of food, shelter and warmth, the material expressions of which are central to archaeological investigations. The Earth in one way or another has furnished these necessities.

Photosynthesizing plants on land and in the oceans extract carbon dioxide from the atmosphere to make tissues that form the base of the various food chains exploited by humans. Oxygen is released in the process. In order to function, however, plants in addition require access to a range of trace elements and minerals. These come ultimately from rocks on land that experience weathering and soil-formation, or an equivalent process on outcrops beneath the sea. In order to help understand the life-styles of human populations and their great movements in the past, trace elements and stable isotopes incorporated from food into human tissues can be used to explore such issues as the nature of people's diets and where people were born and raised. This process requires an appreciation of the broad geological composition of the Earth's continents, as well as an understanding of physiological mechanisms. Geology supplies the necessary petrological and geochemical background to this analysis, allowing a distinction to be made between, say, an inhabitant of a long-stable part of the African crust and a person from a geologically more youthful area in the northern Mediterranean. Geology operates also on a small scale. The diatoms used as environmental indicators, and the phytoliths serving to identify the remains of many food and medicinal plants, incorporate silica into their tissues. Although one of the most stable minerals in the Earth's crust, silica can be slowly leached from soils and so made available to plants.

Geological materials were freely exploited for the earliest shelters that can be dignified by the name of dwellings: sun-dried mud-and-straw bricks for huts thatched with reeds or palms, and post and wattle-and-daub roundhouses. The potential of stone was quickly realized. For millennia rock was levered from natural outcrops or dug from artificial quarries and mines for monument-making, building, and decoration, for making tools, such as querns, millstones, baking stones and whetstones, and for fashioning weapons. This extractive industry continued to be of major economic and social importance up to the present. Geology applied scientifically helps the archaeologist to understand the role of stone in the past. What sorts of stone were involved? Many geological, geochemical, and micropalaeontological techniques are available to help answer this question. From which geological horizons did the rocks come? At which places on the outcrops was the stone extracted and by what means? What properties of the rock made it suitable for the purposes to which it was applied? By carefully mapping the geographical distributions of stone, patterns of trade, and even of authority and polity, can be successfully identified.

The theme of warmth may be broadened from simply the needs of the body to include the use of fire to achieve the moderate temperatures necessary for cooking and the much higher temperatures needed for many technical and industrial operations. Such elevated temperatures are demanded by a wide range of pyrotechnological operations: potting, brick making, tile making, lime-burning, glass-making, the smelting of metals, and the fabrication of metallic implements, appliances, tools, weapons, ornaments and building materials. All but fabricating are underpinned by geology and involve the exploitation of earth-materials extracted from the Earth's crust. Many different clays are known at outcrop but they are not all equally suitable for making pots, bricks or tiles, especially if the intention is to manufacture these items on an industrial scale. Not all limestone formations when burnt produce good lime. The production of metals begins with a search for suitable ores. What do they look like, how do they occur, in which geological settings is it best to search, what is their relationship to the structure of the rocks in which they are found, how is their hidden presence revealed by plant assemblages, and how are they best extracted. The prospector and miner was from prehistoric times onward a practical field geologist applying a lifetime's direct experience to identifying the best sources for the raw materials to meet his immediate needs.

Geological techniques can be used to help confirm where, for example, lime has been burnt or pottery fired in simple clamps. Not only is the soil at these sites likely to be reddened over a substantial area to a shallow depth, but it will also have acquired a distinctive magnetic signature. Sites of metal-working can be detected geochemically by searching for elevated levels of heavy trace-metals, such as copper, lead and zinc. Care is needed, however, in applying this technique. The natural soil or archaeological sediment surrounding the suspect site must be properly sampled and analysed, in order to provide a sound reference base for comparison. It is not sufficient to have found apparently high metal values at the suspect site. Urban living itself introduces heavy metals and elements such as phosphorus to the archaeological environment: the reference base is therefore one that can change over time as occupation continues.

The burning of a carbon fuel is necessary to produce even the modest temperatures required for cooking. Charcoal made by the dry distillation of wood in managed woodlands has been traditionally used for most purposes but coal and coke have also been exploited. Britain at least is very well endowed with both Carboniferous and Jurassic coal-bearing strata. This carbon fuel has been mined since Roman times and underpinned the Industrial Revolution, with its many surviving monuments available for study, including the mines themselves, smelters, foundries, forges and gasworks.

Geology and archaeology thus intersect and are interconnected in many ways. An understanding of geology deepens the appreciation of any archaeological site, be it the context, the artefacts present, or the activities carried out there. The exploration of this interaction, as sketched in this short book, begins with minerals and rocks, the basic components of the Earth's crust, and then considers how they combine to form sequences and structures that can be portrayed on maps, intended to be portable and useful. These depict the intersection of the crustal rocks and the ground surface, both existing in three-dimensions. Adding in knowledge of geological processes, climate, and human activities, the information contained in geological maps should underpin all attempts at understanding landscapes. All landscapes are of archaeological interest, but perhaps the most important, at least in a European context, are those linked to rivers and coasts. As indicated above, mineral extraction is a feature of many landscapes, shown by the presence of pits and quarries. Perhaps the most diverse landscapes associated with mineral extraction are those linked to making metals. They reveal mines, adits, smelters, forges, woodlands managed for charcoal, and complicated routes for the supply of raw materials and product dispersal, such as trackways, paved roads, canals and railways.

2. Minerals

What is a mineral?

The Solar System came into being roughly 4.6 billion years ago. The numerous chemical elements present combined over time with each other in various ways to produce compounds, many of which are found on Earth as naturally-occurring minerals. Acted upon by a range of internal and surface processes, these became the main components of the rocks and sediments that form the Earth's crust. Archaeologists need an appreciation of minerals, for the sediments and rocks from which they are made have long been exploited for building, decorative, milling and sharpening materials, and for the extraction of pigments and cosmetics, gemstones and metals, amongst other items.

A mineral is a substance with a definite and distinctive chemical composition and a regular, repeating atomic structure, the *unit cell*, that determines its physical including optical properties. Common salt – the mineral halite (NaCl) – has a particularly simple unit cell: two atoms of the metal sodium are positioned at opposite corners of each of two opposite faces of a cubic lattice, while four atoms of the gas chlorine occupy the remaining corners (Fig. 2.1A). Minerals possess colour, lustre, transparency/translucency, feel, taste (e.g. rock salt), and odour (e.g. pyrites, FeS_2, when struck). They also have hardness (Moh's Scale), fracture and cleavage. Amongst their most important properties are those of *crystal form* and *symmetry*, the subject of *crystallography*, which may be described by seeking planes and axes about which a good specimen of the mineral is symmetrical.

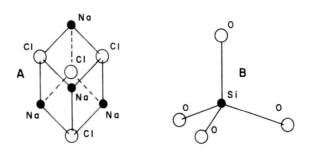

Fig. 2.1 A – Unit cell of common salt (NaCl).
B – The silicate tetrahedron (SiO_4)

Consider a cube, a figure with right-angled corners and six equal faces. It has nine *planes of symmetry* and, passing through a single *centre of symmetry*, thirteen *axes of symmetry*, rotation of the cube about which causes the faces to reappear in the same position (Fig. 2.2). Minerals that form crystals with this high degree of symmetry – critically, they are marked by three equal axes at right-angles intersecting at a single centre – are assigned to the *Cubic System*, of which halite is an example. There are five other Systems, of declining

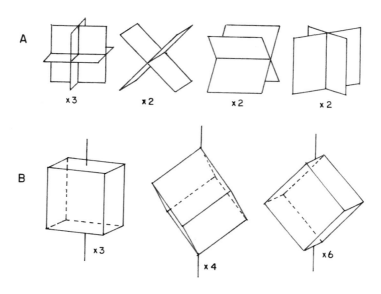

Fig. 2.2 Planes (A) and axes (B) of symmetry of a crystal on the Cubic System. The numerals denote the number of each kind of property

levels of symmetry. *Tetragonal minerals* have two equal horizontal and one vertical axis at right-angles. In *Hexagonal minerals* there are three equal horizontal axes at an angle of 120° to each other and one vertical axis at right-angles. Minerals of the *Orthorhombic System* are typified by three unequal axes all at right-angles. *Monoclinic minerals* are marked by three unequal axes, one vertical, one at right-angles to the vertical axis, and the third at an oblique angle to the plane containing the other two. The common mineral gypsum (see below) is a good example of a monoclinic mineral: its blade-shaped crystals have but one plane of symmetry and

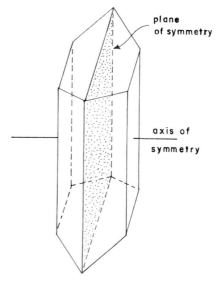

Fig. 2.3 A gypsum crystal showing the single plane and single axis of symmetry

one axis of symmetry (Fig. 2.3). Minerals of the *Triclinic System* have a centre but neither planes nor axes of symmetry. All of these systems are represented among the common rock-forming minerals.

Quartz

Quartz or crystalline silica (SiO_2), a hexagonal mineral, is typically colourless with a glassy lustre and no cleavage (Fig. 2.4A). It is one of the commonest and most abundant minerals on Earth, as any holidaymaker to the beaches of Britain or Ireland can testify. It is a major rock-forming mineral and typical of the light-coloured igneous rocks known as granites. Because of its great hardness and resistance to chemical attack, quartz is the chief constituent of sands and sandstones. Quartz is also found in many mineral veins. Silica occurs as well in cryptocrystalline form as chalcedony, flint/chert – the prehistoric toolmaker's materials of choice– and jasper. Opal is amorphous silica.

Feldspar Family

This chemically diverse group of alumino-silicate minerals is of the highest importance as rock-formers, appearing across the spectrum of igneous rocks. The basic structural unit of the feldspars, as of all silicate minerals, is the SiO_4 grouping of atoms. In this grouping a single silicon atom lies at the centre of a tetrahedral cluster of four oxygens (Fig. 2.1B). The SiO_4 groups in the feldspars are arranged in a framework. The feldspars are colourless, possessed of good cleavage, and often occur in well-shaped, tabular crystals, as in decorative Shap granite. They are much less resistant to mechanical and chemical processes than quartz and so weather readily at the Earth's surface.

One feldspar group is typified by the alkali metal potassium ($KAlSi_2O_8$): *orthoclase* is monoclinic and *microcline* triclinic. These minerals abound in granites. The other important feldspar group, the triclinic *plagioclases*, associates the alkali metal sodium with the alkaline-earth calcium in an isomorphous series of mineral species of gradually changing chemical composition (Fig.2.4B). At the soda-rich end lies *albite* ($NaAlSi_3O_8$) and at the calcium-rich end *anorthite* ($CaAl_2Si_2O_8$). The calcium-rich plagioclases typify the least siliceous, dark-coloured igneous rocks, such as gabbro and basalt.

Olivine Family

The members of this orthorhombic family are important rock-formers that belong to the class known as the *ferromagnesian minerals*. Like the plagioclases, the olivines ($(Mg,Fe)_2SiO_4$) occur as an isomorphous series, the SiO_4 tetrahedra being arranged in regular stacks. The olivine minerals are glassy, without cleavage and, when iron-rich, green, brown or black in hand-specimen (Fig, 2.4C). They typify the least siliceous igneous rocks, such as basalt and gabbro, and are prone to alteration and weathering.

Fig. 2.4 Six common rock-forming minerals in hand-specimen. A – Quartz. (Creative Commons CC-BY-SA-2.5)
B – Plagioclase feldspar. (USGS Licence) C – Olivine. (GNU Free Documentation Licence)
D – Muscovite. (Creative Commons CC-BY-SA-2.5) E. – Calcite. (released public domain image)
F Dolomite. (Creative Commons CC-BY-SA-2.5)

Pyroxene Family

Here is another important class of rock-forming ferromagnesian minerals, appearing chiefly in the dark-coloured, less siliceous igneous rocks. In the pyroxenes the SiO4 tetrahedra are linked to form chains. Two members of this large and varied group are especially important. The orthorhombic pyroxene *hypersthene* $((Mg,Fe)SiO_3)$ is dark-coloured in hand-specimen, with three sets of cleavages. The more common monoclinic *augite*, of variable composition $((Ca,Mg,Fe,Al)_2(Al,Si)_2O_4)$, is greenish black or black in hand-specimen, with a glassy lustre. Again there are three sets of cleavages, two of which intersect at almost a right-angle. The chemical composition of the pyroxenes ensures that they weather readily at the Earth's surface.

Amphibole Family

The most important rock-former in this large ferromagnesian group is *hornblende*, a monoclinic mineral often found as dark green or black, glassy, prismatic crystals with two strong sets of cleavages parallel with the length. The very variable composition sees Ca, Na, Mg, Fe and Al combined in various ways with the $(OH)_2$ radical and SiO_4 tetrahedra in double chains. Hornblende occurs chiefly in moderately siliceous igneous rocks and also in many metamorphic rocks. As with olivines and pyroxenes, the composition ensures that they are prone to alteration and have little resistance to weathering.

Mica Family

The monoclinic micas have many everyday uses and are perhaps amongst the most familiar of minerals because of the single, perfect cleavage displayed: not surprisingly, their component SiO_4 tetrahedra are arranged in sheets. In composition similar to the amphiboles, the hydroxyl radical is invariably also present. Two micas are important as rock-formers. *Muscovite*, or white mica, is a potassium-aluminium silicate found as silvery, tabular crystals with a single perfect cleavage, allowing the mineral to be split into large, thin sheets (Fig. 2.4D). The mineral occurs in many granites and schistose metamorphic rock and also in mineral veins. In the dark green or brown mica known as *biotite*, iron and magnesium are added to the elements found in muscovite. It is otherwise similar in character to muscovite. Biotite occurs in many metamorphic rocks, in some granites, and in many moderately siliceous igneous bodies.

Clay-mineral Family

The members of this diverse and complex family typify muds and mudrocks and are amongst the most abundant in the geological record. They are hydrous aluminium silicates with the invaluable property of becoming plastic when mixed with water and then of retaining their shape when moulded and burnt, characteristics exploited by potters and brick-makers over many millennia. Clay minerals arise quite differently from the minerals so far discussed, which result from either igneous or metamorphic processes within the Earth's crust. Some clay minerals are formed when feldspars and ferromagnesian minerals weather under the moist, oxygen-rich, moderate-temperature conditions found at the Earth's surface. Other clay minerals are the product of soil-forming processes, which also operate superficially. Typically, the clay minerals are flaky and either monoclinic or triclinic, with the SiO_4 tetrahedra in sheets. Their crystals are tiny and generally invisible to the unaided eye.

Kaolinite is a comparatively simple, hydrous alluminium silicate. It is a white material most familiar as the china clay used in potting.

Illite, also known as clay-mica or hydromuscovite, is a hydrous potassium-aluminium silicate, colourless or yellowish-brown. Depending on its origin, illite varies considerably in its degree of crystallinity, which increases as the sediment containing it experiences increasing burial and elevated temperatures. It is the main component of many potting clays.

The name *chlorite* covers a variable group of hydrous silicates of iron, magnesium and aluminium that occur as green, flaky crystals, mainly as the result of the alteration and weathering of ferromagnsian minerals. Some varieties of chlorite include layers of molecular water.

Layers of molecular water are plentiful in the *mixed-layer clay minerals*. These are magnesium-aluminium sililcates that tend to form smaller crystals than the other clays. They are highly desirable in clays for potting and brick-making. Mixed-layer clays are prone to alteration into illite over geological time.

Calcite, aragonite and dolomite

The weathering of rocks at the Earth's surface produces not only particulate debris but also compounds in solution that are carried away by rivers to the seas and oceans. Amongst the most important of these are the salts calcium bicarbonate ($Ca(HCO_3)_2$) and calcium carbonate ($CaCO_3$). In the marine environment, they can be removed in solid form – the bicarbonate indirectly – by chemical processes and especially through the physiological activities of plants (e.g. some algae) and invertebrates (e.g. molluscs, corals) that build hard protective parts or skeletons. The outcomes are the extremely abundant and varied rocks known as limestones. These are important building materials and are also the starting point for the making of lime-mortars by burning.

Calcite, the commonest form of calcium carbonate, is a hexagonal mineral of colourless or white, transparent or opaque, vitreous appearance, with three, very strong cleavages that define a rhombohedron (Fig. 2.4E). These properties, and a simple chemical test, make calcite easy to recognize: a drop of cold, dilute hydrochloric acid placed on the mineral causes it to effervesce. Calcite is also found in lamellar and fibrous varieties, especially in organic materials such as mollusc shells. Calcium carbonate furthermore occurs in an orthorhombic form as needle-like crystals of *aragonite*, as in the skeletons of corals and some other invertebrates. It is readily changed by warmth or pressure into calcite and so, when contributing to carbonate sediments, may not survive burial and lithification.

Calcite is not just the main component of limestones but is also the commonest gangue-mineral in metalliferous ore veins, in which it can appear as large, almost perfect crystals.

Another geologically important carbonate mineral is the hexagonal, double carbonate with magnesium known as *dolomite* (Fig. 2.4F), with the chemical formula $CaMg(CO_3)_2$. Dolomite can be precipitated directly from sea-water to form extensive beds, but mainly arises by the alteration of limestones.

Gypsum and anhydrite

Gypsum is ($CaSO_4.2H_2O$) is the hydrous form of calcium sulphate known as *anhydrite* ($CaSO_4$). As rock-formers, gypsum and anhydrite are classed among the evaporite minerals, for like rock salt they can be directly precipitated from evaporating bodies of sea-water under hot climatic conditions. Gypsum is a colourless, monoclinic mineral (Fig. 2.3) marked by its single, strong cleavage, softness, and greasy feel. It can be found as small clusters of radiating crystals or as large, commonly twinned, blade-like forms. When burnt so as to drive off the combined water, gypsum is converted to plaster of Paris, the basis of special mortars and plasters used in building.

Another origin for gypsum is through the weathering in the presence of calcareous material of pyrite (FeS_2) associated with decaying plant material, as in organic-rich muds and mudrocks during the early stages of burial. The mineral in this case is of secondary origin.

Minor and accessory minerals

In addition to representatives from amongst the common rock-formers, igneous and metamorphic rocks contain other minerals in minor or accessory amounts. These subordinate components are known as 'heavy' minerals, because their density is generally much greater than that of the major components. On weathering, metamorphic and igneous rocks release heavy minerals in small amounts into the sands and muds that will become sedimentary deposits, from which they can be extracted by separating them off in a suitable heavy liquid, such as aqueous sodium tungstate. Heavy minerals have proved to be valuable markers for correlation and indicators of provenance. An early archaeological application was the resolution of the long-debated source of the Romano-British coarse ware known as Southeast Dorset Black-burnished Ware No. I. More recently they have been used to settle the provenance of building stone used in Roman times in coastal Norfolk and the sandstones for making whetstones in the Weald.

The commonest accessory minerals of granitic rocks are *apatite* ($Ca_5F(PO_4)_3$), *rutile* (TiO_2), *sphene* ($CaTiSiO_5$), *topaz* ($Al_2F_2SO_4$), *tourmaline* (a complex aluminium borosilicate), and *zircon* ($ZrSiO_4$).

Metamorphic rocks afford a different, characteristic suite of minor and accessory minerals of generally complex composition and structure. These are *chlorite* (complex aluminium-iron-magnesium silicate), *epidote* (hydrous aluminium-calcium-iron silicate), *garnet* (aluminium-iron silicate), *kyanite* (Al_2SiO_5, triclinic), *sillimanite* (Al_2SiO_5, orthorhombic) and *staurolite* (hydrous aluminium-iron silicate).

Further reading

Gribble, C.D. 1988. *Rutley's Elements of Mineralogy*, 27th ed., Berlin, Springer
MacKenzie, W.S. & Adams, A.E. 1994. *Rocks and Minerals in Thin Section*. Boca Raton, CRC Press
Mange, M.A. & Maurer, F.H.W. 1992. *Heavy minerals in colour*. London, Chapman & Hall
Vaughan, D.J. 2014. *Minerals: a very short Introduction*. Oxford, OUP

3. Rocks and Sediments

Major classes

Rocks and sediments are aggregations of minerals that have come into being as the result of natural processes. They are of three broad kinds: *igneous*, *sedimentary* and *metamorphic*.

Igneous, or fire-formed, rocks depend on the Earth's internal sources of heat and originate within the deeper crust as molten material or *magmas*, which after emplacement upward eventually cool, crystalize, and solidify. The eruption of volcanoes brings magmas to the Earth's surface in the form of lavas and ashes. Under certain circumstances, although of igneous origin, volcanic ashes can become involved in Earth-surface processes. Accordingly, they provide a bridge to sediments.

Sedimentary deposits result from processes that operate at the Earth's surface under the low-temperature, oxygen-rich and generally moist conditions that prevail there. Rock weathering at the surface gives rise to broken, particulate debris and also to dissolved substance. These can all be fashioned by processes of transport and deposition into layers of sediment and then, as the result of compaction and cementation during burial, into sedimentary rocks.

Metamorphic rocks are as diverse as igneous and sedimentary rocks put together, because they arise when these parent materials become buried, heated and subjected to shearing forces deep in the Earth's crust.

How to study rocks

Fieldwork has primacy in geology as in archaeology. Rocks are best first studied in the field, where their character and relationships can most clearly be established. Rocks display structures in the form of joints and varieties of bedding. As materials, they can be examined in the field using a hand-lens – an indispensible tool – on a fresh or clean, naturally weathered surfaces, in order to ascertain their mineral composition and such features as crystal or grain size. The next step up is to prepare from a hand-specimen in the laboratory a thin-section of the rock, at the standard thickness of 30 μm, for examination under a petrographic microscope. This is especially important in the case of fine-grained rocks with components difficult to resolve otherwise. Minerals can also be identified using x-ray diffraction techniques, a valuable tool where fine-grained materials are concerned. Many laboratory methods of chemical analysis can be applied to rocks and sediments, from the cheap-and-cheerful, such as x-ray fluorescence analysis, to the precise and costly. It is now possible to determine bulk chemical compositions in the field using a portable x-ray fluroescence analyser but the technique can prove unreliable.

Igneous rocks

The crystallizing out of minerals as magmas chill results in the texture so typical of igneous rocks. The slower the rate of cooling the larger the crystals which form. Glassy lavas like obsidian and pitchstone, however, have chilled so rapidly that crystals could barely begin to emerge. Because the chemical composition of magmas is generally complex, some minerals crystallize out early on and others at a later stage. Early crystals tend to be large, well-shaped and sometimes flow-oriented. They are known as phenocrysts. Later minerals, infilling what space remains, are seldom perfect in form.

Igneous rocks are classified from *acid* to *basic /ultrabasic* on the basis of their silica content. Igneous bodies take a variety of forms, depending on how and where in the Earth's crust they were emplaced.

Granite is a widespread and common acid igneous rock in Britain and Ireland. They are light-coloured and consist of abundant quartz and potassium feldspar, often in the form of phenocrsysts, with subordinate micas (Fig. 3.1). Granites occur in huge masses called plutons that were intruded at a depth of some kilometres into the crust. Geophysical work has shown that the granite outcrops of the Southwest of England – Dartmoor, Bodmin, Hensbarrow, Carnmenellis, Land's End – are merely the uppermost parts of a vast *pluton* that underlies almost the whole peninsula. The final stages in the cooling of granites commonly sees the creation of mineral veins with valuable metallic ores. *Rhyolite* and related lavas and ashes are the fine-grained, extrusive equivalent of granites. They are particularly associated with volcanoes that erupt explosively

At the opposite end in the range of chemical composition are dark-coloured, basic igneous rocks such as *gabbro* and *basalt,* typified by low silica and no free quartz. Gabbros are coarse-grained intrusive rocks formed typically of enmeshed plagioclase feldspar crystals with subordinate, interstitial ferromagnesian minerals such as pyroxene and olivine (Fig. 3.2A). The equivalent lava is basalt (Fig. 3.2B), formed from low-viscosity magmas that flood large areas around relatively flat, shield or fissure volcanoes marked by little explosive activity. Basalts underlie the 'trap' landscapes of Mull, Skye and Northern Ireland. *Dolerite* is a basic rock of intermediate crystal size commonly found in the form of sub-horizontal, sheet-like *sills* (e.g. Whin Sill, northeast England) or narrow, near-vertical *dykes.*

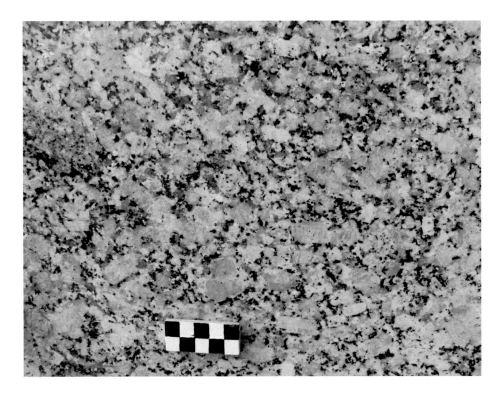

Fig. 3.1 A megacrystic granite with large, flow-oriented feldspars.. Scale 5 cm

Lying compositionally between acid and basic rocks are those of *intermediate* composition, such as the intrusive *syenites* and *diorites*. They consist of mixed potassium and plagioclase feldspars, some quartz, and hydrous ferromagnesian minerals such as hornblende and biotite. An important, equivalent lava is *andesite*.

Clastic sediments and rocks

Rocks exposed at the Earth's surface encounter physical and chemical conditions greatly different from those under which they were formed. They are consequently prone to *weather*. *Physical weathering* leads to the disintegration of the rock. It is effected by freeze-thaw, as water trapped in pore spaces, joints and partings freezes, expands and thaws (cold/cool climates), and by the pressure exerted by salts as they crystallize out from entrapped water (warm/hot climates). Root growth and heating-cooling also aid physical weathering. Physical weathering affords broken and fragmentary materials, whence the appellation *clastic* when they combine to form sediments. The debris yielded by physical weathering ranges hugely in particle size from *gravel* (>4 mm), through *granules* (2-4 mm) and *sand* (63 μm-2000 μm), to *mud* (silt and clay, 2 μm-63 μm).

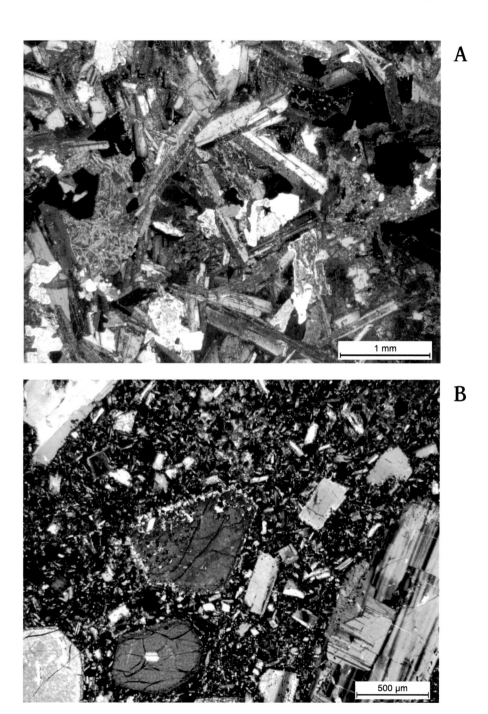

Fig. 3.2 Selected rocks in thin-section. A – gabbro, doubly-polarized light . B – basalt, doubly-polarized light. C – calcite-cemented fine-grained sandstone, Weald Clay Formation (Early Cretaceous), plain-polarized light. D – lithic sandstone (Pennant Measures, Upper Carboniferous). E. – shelly limestone (Upper Jurassic), plain-polarized light. F – oolitic limestone (Middle Jurassic), plain-polarized light

C

500 μm

D

500 μm

Fig. 3.2 continued. Selected rocks in thin-section. A – gabbro, doubly-polarized light . B – basalt, doubly-polarized light. C – calcite-cemented fine-grained sandstone, Weald Clay Formation (Early Cretaceous), plain-polarized light. D – lithic sandstone (Pennant Measures, Upper Carboniferous). E. – shelly limestone (Upper Jurassic), plain-polarized light. F – oolitic limestone (Middle Jurassic), plain-polarized light

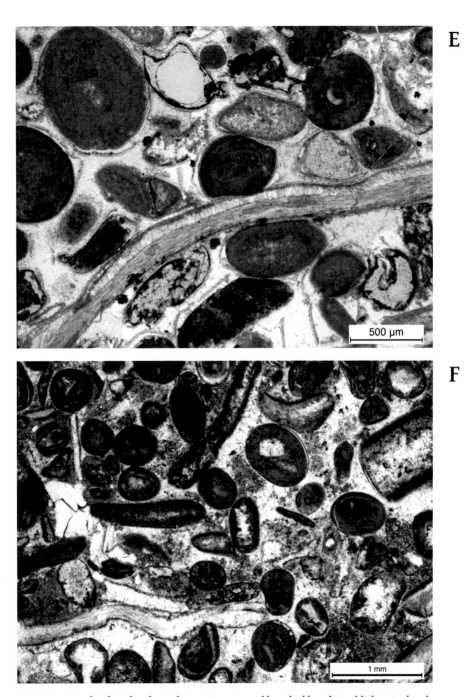

Fig. 3.2 continued. Selected rocks in thin-section. A – gabbro, doubly-polarized light . B – basalt, doubly-polarized light. C – calcite-cemented fine-grained sandstone, Weald Clay Formation (Early Cretaceous), plain-polarized light. D – lithic sandstone (Pennant Measures, Upper Carboniferous). E. – shelly limestone (Upper Jurassic), plain-polarized light. F – oolitic limestone (Middle Jurassic), plain-polarized light

500 µm

Fig. 3.3 A Cretaceous glauconitic sandstone (doubly-polarized light)

By opening the pores and fissures present in rocks, physical weathering fosters *chemical weathering*, that is, the rotting of especially the feldspar and ferromagnesian minerals, particularly where temperatures are high and rainfall plentiful. New particulates, notably clay minerals, are formed along with dissolved substances that are carried away by rivers and groundwater, and eventually reach lakes and oceans.

Because the particles are so large, gravels arise only where water currents are very strong, such as on alluvial fans, in swift rivers and on exposed beaches. When buried and cemented to form a rock, gravels become *breccias*, if the particles are angular, or *conglomerates* when substantial particle rounding has been achieved through transport. Gravel particles on river beds or beaches often show a preferred orientation, or shingling, due to current action.

Sands are amongst the commonest sediments, to be found in rivers, on beaches, and in the desert. They are highly porous and permeable. When buried in the crust and cemented sands lose much if not all of their porosity and give rise to *sandstones*. The typical sandstone consists of two contrasted components. The original sand – the *detrital* element – is normally a mixture of minerals and mineral-aggregates. Quartz is commonly dominant and may be the only mineral to form the sand. Feldspars may also be present, along with glauconite and comminuted fossil debris (Fig 3.2C). Some sandstones are rich in fragments of fine-grained rocks, such as slate or lava, what are described as *rock fragments* or *lithic debris.* (Fig. 3.2D). To these detrital components is normally added during burial and lithification a post-depositional cement, typically either calcite or quartz in optical continuity with primary grains. Sandstones cemented by secondary quartz are known as *quartzites* and those rich in feldspars as *arkoses.* Some sands are so rich in rock fragments that, when buried, all porosity is lost as the grains are tightly squeezed together and deformed (Fig. 3.2D). In the British geological record, there are distinctive green sandstones coloured by the iron-rich mineral *glauconite*, a product of warm-water conditions (Fig. 3.3).

Sands and sandstones record environments where transport by wind was possible, in deserts and on exposed coasts, and where water currents attained moderate speeds, such as river channels, deltas, tidal estuaries, wave-swept beaches and shallow tidal seas. The action of currents during the transport of sand imprints distinctive shapes on the surface of the deposit and patterns of lamination within the interior. These are known as *sedimentary structures*, many of which hint at current strength and direction. Common examples are *ripple marks* seen on bedding surfaces (Fig. 3.4) and *cross-bedding* found internally as horizontal stacks of steeply tilted laminae

Fig. 3.4 Current ripples on a modern beach. Flow towards upper right. Pocket tape c. 5cm square

Fig. 3.5 Cross-bedded Upper Carboniferous sandstone. Flow from left

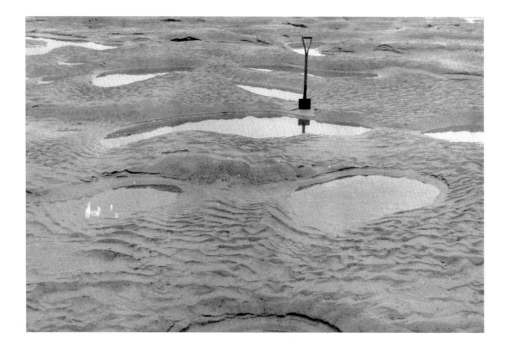

Fig. 3.6 Dunes formed by tidal currents in a modern estuary. Flow towards reader

(Fig. 3.5). Cross-bedding, common in aeolian, tidal and fluvial sandstones, is formed as the large, asymmetrical mounds of sand called *dunes* are driven forward by the wind or flow of water (Fig. 3.6). Dunes migrate downcurrent as grains scoured from the gentle, upstream face settle to form laminae on the steeper, downstream side. On beaches, the action of the surprisingly strong currents created by wave swash and backwash gives rise to another but sub-horizontal pattern of bedding, known as *parallel lamination*. Flaggy sandstones are typified by the presence of this particular structure (Fig. 3.7).

Muds and *mudrocks* are the most voluminous sediments in the geological record and amongst the most important economically, as the basic material for making ceramics. These sediments have formed on river floodplains, coastal marshes, in estuaries and deltas, and especially in seas distant from the influence of land. Mud is a complex mixture of clay minerals, quartz and feldspar silt, and particulate to semicolloidal vegetable matter of coastal or terrestrial origin. In saline environments it is normally loosely flocculated. Mud is sufficiently fine-grained as to be transported in suspension by water, settling out only under calm conditions, such as at high tide in an estuary. Freshly deposited mud has a high water content, running in many cases to 60-70%. Burial squeezes out the water, greatly reducing the thickness of the deposit, and rotates the largely platy clay minerals to a mainly sub-horizontal attitude. Fissile *shales* are the mudrocks expressive of this effect.

Glacial deposits and features

Ice-sheets and glaciers as they move scour their rocky beds, entraining debris ranging in size from small fractions of a millimetre (rock flour) up to blocks and rafts as big as houses. When as the result of climate change the ice stops moving, that is, during *deglaciation*, this debris is released to form a range of *glacial deposits*. Such sediments are widespread in Britain and Ireland. They strongly influence soil-formation and land-use and some are economically important.

The sediments directly deposited from ice are known as *tills* (old name boulder clay). Basal or *subglacial tills* accumulate at the base of the ice as it melts from below. Typically, these tills are strong, compact, unbedded, and extremely ill-sorted. The little internal stratification occasionally to be found reflects either zones of shearing within the ice or deposition from subglacial streams. On the other hand, *supraglacial tills* accumulate on the uneven surface of the melting ice and at the ice-front. They commonly display some bedding, as the result of mass-movement processes and meltwater runoff, and tend to underlie irregular ridges and mounds (*moraines*).

Meltwater carrying sediment flows over and also beneath the surface of shrinking ice-sheets and glaciers, depositing *glaciofluvial* sands and gravels on the margins of the ice and in tunnels within. The tunnel deposits, when deglaciation is complete, are seen as ridges, known as *eskers*, which snake through the landscape, making on a regional scale dendritic patterns like river networks. Most of the sediment released is deposited beyond the ice by meltwater as extensive sheets of *proglacial outwash*. These well-bedded deposits have all the sedimentary structures expected of normal river sediments, but they can include masses of sand incorporated during the spring melt as frozen blocks.

At high latitudes and around the margins of ice sheets the ground is permanently and deeply frozen (*permafrost*), except for the topmost few metres subject to seasonal freeze-thaw (*active zone*). This is the *periglacial zone*. Here are formed several features and structures – *patterned ground*, *ice-wedge casts*, and *involutions* – that have often confused archaeologists.

Carbonate sediments and rocks

Calcium carbonate dominates an important group of mainly marine rocks and sediments. The dissolved mineral is removed from sea-water in two ways, through the physiological activities of invertebrates and phytoplankton, and by direct chemical precipitation.

Many marine invertebrates, such as molluscs and corals, secrete calcite or aragonite in order to make hard parts. On the death of the organisms, these hard tissues can be swept up and sorted by wave or tidal currents into beds of biogenic detrital sand or gravel which, after cementation by more calcite, become *shelly limestones* (Fig. 3.2.E). Many of the sedimentary structures familiar from terrigenous sandstones are to be found in these rocks.

The single-celled algae called coccolithophores, an important, bloom-forming element of the marine phytoplankton, build spheroidal carapaces of tiny, oval crystals of calcite. After death, these sink to the sea-bed to make a variety of white calcite mud that on lithification becomes the familiar *chalk*, or chalkstone, in which flint can be precipitated during early burial.

The mild evaporation of sea-water under warm climatic conditions commonly leads to the supersaturation of dissolved calcium carbonate, whence direct chemical precipitation may follow. In calm waters far from land white clouds of aragonite needles may form, settling to the sea-bed to make another variety of carbonate mud. In clear, agitated, and especially tidal waters onion-like layers of calciuim carbonate can be precipitated around nuclei – fragments of shell or grains of quartz – to make egg-like ooids. These accumulate as oolite shoals under the action of currents, eventually to lithify as *oolitic limestones* (Fig. 3.2F). Ripple marks and cross-bedding are commonly preserved in them.

Evaporites

The severe evaporation of sea-water under hot and dry climatic conditions causes other dissolved minerals to precipitate out in a rough sequence, notably gypsum ($CaSO_4.2H_2O$), then halite (NaCl) and finally polyhalite ($K_2SO_4.MgSO_4.2CaSO_4.2H_2O$). These mineral deposits, called *evaporites*, occur as thick beds of great economic importance amongst the Permian and Triassic rocks of Britain. Brines derived from beds of rock salt have been exploited since before Roman times.

Metamorphic rocks

These are igneous and sedimentary rocks that have been transformed as the result of being forced deep into the Earth's crust. The elevated temperatures there cause mineral recrystallization and even partial melting. The softened rocks become squeezed and sheared as the result of the high pressures due to the overburden and the compressive forces acting within the crust. Little or no bulk chemical change occurs during metamorphism, but new mineral assemblages, and new textures, and commonly flow-like structures normally result. These associations typify the particular conditions or *grade* of metamorphism experienced, for example, the temperature to which the parent rocks were heated.

Fissile *slate* used for roofing is one of the most familiar of metamorphic rocks (Fig. 3.8). Typified by the presence of cleavage, its origin lies in mudrocks that were heated and squeezed to such an extent that the original clay minerals recrystallized to illite aligned perpendicular to the direction of the acting horizontal stress. Slate is a fine-grained, low-grade metamorphic rock.

Schists represent a moderate grade of metamorphism. These are medium- to coarse-grained, foliated rocks, consisting mainly of well-crystallized quartz and often very abundant muscovite with albite feldspar. Some schists are rich in biotite and others in chlorite or garnet. When a quartz-rich sandstone is metamorphosed at this grade it become a *metaquartzite*, in which the quartz assumes the form of stretched, sutured crystals.

The highest grades of metamorphism are represented by *gneisses* and *migmaties*. The typical gneiss is medium- or coarse-grained and composed of well-crystallized quartz with albite and micas and commonly garnet. Gneisses display a strong, flow-like foliation due to the presence of bands of minerals of different composition (Fig. 3.9).

Whereas mudrocks and sandstones, and combinations of the two, lead to schists, gneisses and migmatites on metamorphism, limestones give rise to *marbles*

Fig. 3.7 Flaggy (parallel-laminated) Upper Carboniferous sandstone

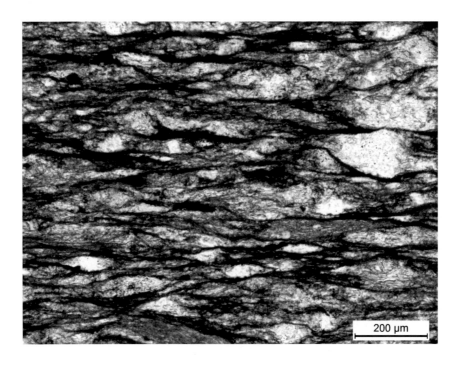

Fig. 3.8 A coarse, silty slate in thin-section (plain-polarized light)

Fig. 3.9 A banded garnet-gneiss. Scale 5 cm

in which the calcite is recrystallized, often quite coarsely. Many marbles are white or palest grey, but others are attractively coloured and patterned, due to the new minerals created from the insoluble components of the parent rocks. Marbles are an important decorative stone.

Catastrophic deposits

Some Earth-surface processes and agencies act with extreme violence and over very short time-scales, and accordingly may be described as *catastrophic*. Examples are rock-falls, avalanches, landslides, flash floods, explosive volcanicity, storm surges and tsunamis. Amongst the most important archaeologically are *explosive volcanicity* and *tsunamis*.

Volcanoes fed with acidic magma tend to explosive behavior, as archaeologists familiar with the fate of Pompeii and Herculaneum will recall. Gas-charged magma may burst from the upper slopes of the volcano or be violently forced upward from the crater into the atmosphere as a towering eruption column. After collapsing onto the slopes of the volcano, the discharged material may rapidly surge downward as a dense cloud of hot gas, crystals and foam-like, disintegrated magma. There results an ill-sorted, chaotic deposit, but some structures indicative of flow may be presemt.   The finest ash, suspended in the atmosphere, may fall as a smothering blanket over large parts of the surrounding countryside.

Earthquakes centred on the edges of the ocean may cause the collapse of huge volumes of the sediment beneath the ocean floor. The resulting sudden deepening of the water creates a fast-moving, solitary surface wave capable of travelling for hundreds of kilometres away from the site of the earthquake; similar effects may arise as the result of volcanic explosions at sea. When such a wave enters shallow water close to land, it steepens up to form a tsunami many metres in height that is capable of travelling inland for many kilometres, drowning coastal lowlands and surmounting the lower hills. Tsunami deposits are also ill-sorted, chaotic, and sometimes bouldery. Their organic content is a distinctive mixture of marine organisms and terrestrial plant matter.

Further Reading

Adams, A.E., MacKenzie, W.D. & Guilford, C. 1984.. *Atlas of Sedimentary Rocks under the Microscope*, Harlow, Longman

Jones, A.P., Tucker, ME. & Hart, J.K. 1999. *Description and Analysis of Quaternary Stratigraphic Field Sections.* London, Quaternary Research Association

MacKenzie, W.S., Donaldson, C.H. & Guilford, C. 1982. *Atlas of Igneous Rocks and their Textures.* Harlow, Prentice –Hall

Walkden, G. 2015. Finding our marbles. *Geoscientist* 25(4), 10-15

Yardley, B.W.D., MacKenzie, W.S. & Guilford, C. 1990. *Atlas of Metamorphic Rocks and their Textures.* Harlow, Longman

4. Geological Maps

What is a geological map?

The bodies of rock that make up the Earth's crust exist in three-dimensions, with time regarded as a fourth. Most are stratified – the sedimentary deposits – but many, especially the intrusive igneous rocks, take a range of other forms, depending on their geological age and origin. The shape of the Earth's surface, its topography, is also three-dimensional. The art of the geological map-maker is to portray, on two-dimensional paper, the intersection of these three-dimensional forms, that is, the outcrops and structure of the rocks and sediments as they appear at the surface. Because of developments in computing, it is now possible to assess traditional geological maps in the form of three-dimensional images capable of being edited and rotated.

The first geological map of real significance was made by William 'Strata' Smith (1769–1839), an Oxfordshire land and mine surveyor who made his living by advising on the construction and operation of canals, coal mines, drainage schemes, sea-defences, and harbours. Smith was active over many decades and travelled extensively throughout England and southern Scotland in connection with his work. He came to realize that the mainly stratified rocks he encountered in cuttings, mines, river-beds and cliffs could be ordered in a definite upward sequence and successfully correlated from place to place, especially by means of the fossils they yielded. The outcome was the national geological map he published in 1815, covering England, Wales and the Southern Uplands and Lowlands of Scotland (Fig. 4.1). This depiction of outcrops was an extraordinary, if underappreciated, achievement for its time and has proved to be remarkably accurate.

How geological maps are made

The archaeologist employs in the field a distinctive set of simple tools: spade, mattock, trowel, pan and brush, measuring tape, planning frame and surveying equipment. The tools used by the field geologist when interrogating the Earth – a topographic map (typically 1:10,000 scale), magnetic compass, clinometer, measuring tape, hammer, hand-lens, acid-bottle, notebook, and sometimes an auger – are equally simple and not that different. As in archaeology, air photographs are an invaluable aid. What is the evidence from which a geological map can be constructed using these tools?

Fig. 4.1 William Smith's geological map of 1815. The Carboniferous rocks of the Mendips, South Wales and the Pennines appear in dark blue and grey. The Jurassic and Cretaceous rocks are shown in bright red, yellow and dull greyish green. Note the granites of Southwest England depicted in bright red. Copyright: British Geological Survey

The most direct evidence takes the form of *exposures* of rock on hillside crags, in streams and quarries, and on cliffs. Use of the hand-lens and acid bottle on fresh specimens allows the kinds of rock present to be established. If the rocks are bedded, the attitude of their *bedding planes* – the extent of their post-depositional disturbance expressed by the value and direction (azimuth) of their maximum dip – can be established and recorded using the compass and clinometer. The tape measure allows a *log* to be made through a sequence of beds as they follow each other in order of deposition. Once these tasks are accomplished at an exposure, and the details entered in the notebnook, the site is plotted on the topographic map and a summary of the lithology and dip of the beds added.

More subtle means must also be used when mapping in Britain and Ireland, where the rocks are largely concealed beneath soil and vegetation. The colour and texture of the *soil* itself can provide important clues as to what lies beneath, especially where ploughing has brought up fragments of rock from the subsurface, known as 'brash' or 'float'. A similar effect arises where rabbits or badgers have dug burrows or sets. It may otherwise be necessary to auger into the soil. The sites of auger holes and burrows can be plotted and the fields printed on the map annotated with the character of the float. Guidance may also come from plants that are specific to particular kinds of underlying rock, and from the locations of seeps and springs at faults or where permeable beds overlie impermeable ones. A distinctive flora grows on calcareous soils. Bracken is a good indicator of well-drained gravels and sands.

The final clues come from what Williams Smith called 'the lie of the land', for which he had an experienced and well-developed eye, a skill that archaeologists also do well to cultivate. As rocks weather at the Earth's surface, the harder and more resistant ones give rise to hills and, if they are bedded, to ribs, ridges, and escarpments that can be walked out and mapped across the land. In such terrain the softer beds weather as receding hollows. The result is a distinctive *dip and scarp* topography (Fig. 4.2) that in some districts can be traced out for many kilometres, as in the Pennine coalfields and the lava-related 'trap' landscapes of Mull and Skye in the Inner Hebrides and the Antrim Plateau of Northern Ireland. Occasional crags on the ribs and hills allow the nature of these feature-forming rocks to be confirmed. Developed on a large scale, as in the Chilterns and Cotswolds of southern England, this type of landscape offers features conspicuous and extensive enough to be chosen by people to define trading areas and polities. Level terraces with scarps are another kind of feature of mapping significance. These are commonest in river valleys (e.g. Thames) and in the coastal zone, where they express stages in the development of the landscape by marine processes.

The geologist setting out to make a map first gathers and plots the above kinds of evidence as found in a small, initial portion of the topographic sheet. While in the field, boundaries between the various rock units are recorded using the evidence, any breaks in the strata marked as faults, and an explanatory column begun as a key to the sequence of formations. Proceeding as before, the boundaries between the bodies of rock are followed out until the whole of the sheet has been covered. The explanatory column is then revised, the formations named and coloured in, and one or more explanatory cross-sections drawn up to illustrate the geological structure of the area. The map is then complete and can be used to answer a variety of practical questions.

Because they have a topographic base (albeit an uncoloured one, avoiding confusion), the significance for archaeologists of geological maps is that they point up the context for human settlement, land-use, industrialization, trading, and military activities. In the case of landscapes of alluvial origin – river valleys and many coastal areas – geological maps merge with geomorphological ones in revealing the successive stages in their development. In Britain, geomorphology and geology have unfortunately tended to develop as separate disciplines.

Faults

Earth movements in response to internal forces over the course of geological time have fractured the essentially brittle rocks that form the upper part of the Earth's crust. These fractures, known as *faults*, are very common and are shown on geological maps as sudden breaks and displacements in the lateral continuity of the strata, marked by heavy lines or heavy, broken lines. Faults disappear with depth, either fading out or joining other faults.

Faults are of four main kinds. *Normal faults* arise when the rocks, acted on by horizontal tensional forces, break as they become stretched (Fig. 4.3A). The largest component of movement on these faults is vertical. The largest component of movement on *reverse faults* is also vertical (Fig. 4.3B), but in this case the beds are affected by horizontal compressive forces and so become shortened. A second type of shortening fault is the *thrust fault*, arising under a heavy overburden of strata (Fig. 4.3C). The main movement in this case is subhorizontal and the fault plane lies at a low angle, in contrast to the reverse fault. The largest thrust faults, to be found in the Scottish Highlands, for example, can displace strata for many kilometres. The *strike-slip fault* also leads to a horizontal displacement but has a generally steep fault plane (Fig. 4.3D). The largest strike-slip fault in Britain is the Great Glen Fault in the Scottish Highlands, over 100 km long and with a proven final horizontal displacement of 64 km. Active over a long period of geological time, it determines the location of the eponymous topographic feature that now contains a main road and an impressive ship canal linking the east and west coasts.

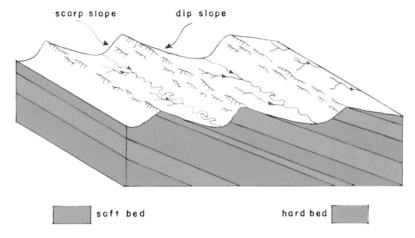

Fig. 4.2 Schematic dip-and-scarp topography

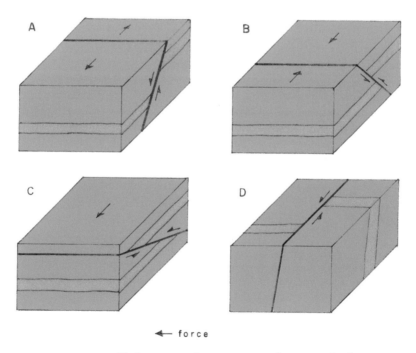

Fig. 4.3 Types of fault. A – Normal. B – Reverse. C – Thrust. D Strike-slip

Folds

On a geological time-scale, strata covered by a sufficient overburden can respond in an effectively plastic manner to horizontal compressive forces, and so become shortened as the result of *buckling* or *folding*. Folds are of four main kinds, each with a distinctive outcrop pattern.

An *anticline* (Fig. 4.4A) is an upright fold that is broadly symmetrical about a near-vertical axial surface with the older beds in the core and the younger ones in the outer limbs. They have a distinctive appearance from the air or in a satellite image, especially if alternately hard and soft beds are present to make a dip-and-scarp topography. Anticlines vary considerably in scale. Amongst the smaller in Britain are the Shalbourne and Kingsclere Anticlines in north Hampshire. One of the largest, alternatively described as a dome, is to be found in southeast England, stretching eastward from Hampshire through Surrey and Sussex to Kent and the Strait of Dover. With bold, inward-facing scarps on the northern and southern limbs, formed by the North and South Downs, it is clearly evident on William Smith's map (Fig. 4.1).

The *syncline* (Fig. 4.4B) is the opposite of an anticline, the limbs now dipping inward toward the axial surface and any scarps facing outwards. In the real world synclines, like anticlines, tend to die out or 'plunge' along their length. They combine with anticlines to make *fold-belts*, such as the major one that can be traced through the Mendips and coastal South Wales.

The *pericline* (Fig. 4.5A) is a type of an unsymmetrical fold with just one, near-vertical limb. Periclines are generally large but not common structures. Examples are to be found in Somerset (Mendips), southeast Dorset (Isle of Purbeck), and the Isle of Wight. The latter pericline is sufficiently large as also to be shown on Smith's map (Fig.4.1). The Dorset pericline is famous for its mines and quarries, favoured by the steep dip, dug for Purbeck marble, a decorative stone much favoured since Roman times.

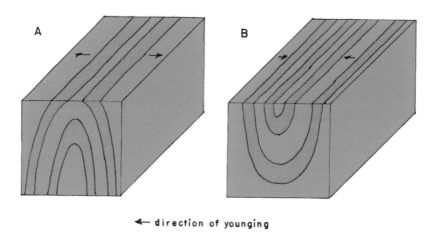

← direction of younging

Fig. 4.4 Types of fold. A - Anticline. B - Syncline

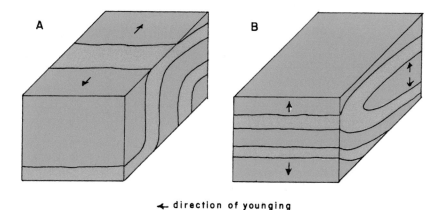

Fig. 4.5 Types of fold. A - Pericline. B - Recumbent (overfold)

The *recumbent fold* (Fig. 4.4B) has a flat-lying axial surface and represents an exceptional degree of horizontal shortening of the crustal rocks. They are commonly associated with thrust faults. The largest recumbent folds are known as *nappes,* examples of which occur in the Scottish Highlands.

Unconformities

Often styled 'The Father of Modern Geology', James Hutton (1726–1797) was born and educated in Edinburgh where he trained as a physician, although he also farmed and was a keen naturalist. He earned this appellation because he grasped, as the result of his geological wanderings, that the Earth had a long history in deep time, and that the Earth's crust and surface had repeatedly been renewed. His main evidence was the *unconformity* (Fig. 4.6), a sharp break

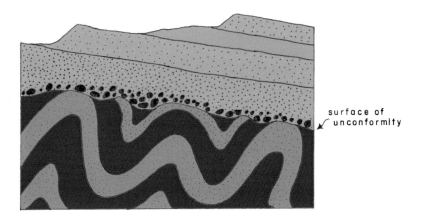

Fig. 4.6 A schematic unconformity in vertical, two-dimensional section

in the sequence of rocks typically revealed by an abrupt change of structure and lithological character, pointing to a period of uplift of the crust and the weathering and erosion of the beds thus exposed. Figure 4.7 shows one of the several unconformities discovered by Hutton.

An unconformity means that the deposition of sediment at a location was brought to a halt by earth movements and that, for a period of time, commonly running into millions of years, uplift and deformation and erosion took its place before deposition was resumed. The archaeological equivalent is the break in a sequence of archaeological deposits at a site marking the interval before the site was reoccupied. A good example is an extensive burnt layer.

National Geological Surveys and other data sources

Where can the archaeologist obtain relevant geological information about Britain and Ireland? For strategic reasons, the state founded geological surveys in Britain and Ireland during the nineteenth century as the countries became increasingly populous and industrialized. The *British Geological Survey*, established in 1845 as the Geological Survey of Great Britain, now covers England, Wales and Scotland. This agency publishes, in paper and digital form, a national coverage of geological maps in numbered, coloured sheets at a range of scales, of which the 1:50,000 is the standard. These maps, frequently revised, are accompanied by either an explanatory *memoir* or by a *summary explanation* printed on the map itself. The older memoirs are especially valuable because they include local details of exposures, such as small quarries, that have long disappeared from the ground. The more specialized publications cover offshore areas, geophysical and geochemical data, resources such as stone, water, sand and gravel, and metalliferous deposits.

Northern Ireland has its own geological survey, the *Geological Survey of Northern Ireland*. Its publications are similar in character and scope to those of the British Geological Survey.

Geological knowledge of the Republic of Ireland is the responsibility of the *Geological Survey of Ireland* (1845). It produces maps, memoirs and specialist reports similar to those of the British Geological Survey. A complete national coverage of geological maps and memoirs was an early achievement of this Survey in the nineteenth century.

The above agencies produce 'official', systematic knowledge about the geology and structure of Britain and Ireland, but they are far from being the only sources for the needful archaeologist. The publications of learned and semi-

Fig. 4.7 Hutton's famous unconformity at Siccar Point on the northeast coast of the Southern Uplands of Scotland. Near-vertical Silurian sandstones and mudstones are overlain by now-tilted beds of Old Red Sandstone (Devonian). Image by Dave Souza – Creative Commons Licence 2.0

amateur societies also have importance. In England and Wales, they include the *Philosophical Transactions of the Royal Society*, the *Journal of the Geological Society* (of London), the *Geological Magazine*, the *Proceedings of the Geologist's Association*, the *Proceedings of the Yorkshire Geological Society*, and the *Geological Journal*. Scotland has the *Transactions of the Royal Society of Edinburgh*, and the two Irelands the *Proceedings of the Royal Irish Academy*, the *Journal of Earth Sciences of the Royal Dublin Society* and the *Irish Naturalist*. There are many specialized journals. The *Transactions of the Institute of Mining and Metallurgy* covers all five countries. Of considerable value are international journals such as *The Holocene*, the *Journal of Quaternary Science*, the *Journal of Archaeological Science*, and *Quaternary Science Reviews*. It is worth noting that many county field clubs publish journals, often with important geological and historical/archaeological contents.

Further Reading

Allen, P.M. 2003. *A geological survey in transition*. Nottingham, British Geological Survey

Bennison, G.M. 1990. *An introduction to geological structures and maps*, 5th. ed. London, Edward Arnold

Maltman, A 1998. *Geological maps: an introduction*, 2nd. ed. Wiley, Chichester

McIntyre, D.B. & McKirdy, A. 1997. *James Hutton: the founder of modern geology.* Edinburgh, The Stationery Office

Winchester, S. 2001. *The map that changed the world: the tale of William Smith and the birth of a science.* London, Viking

5. Geological Stratigraphy

Succession and correlation

Noting the way in which sediments built up in ditches, ponds and lakes, geologists such as James Hutton and William Smith grasped early on that, with some exceptions, younger strata came to overlie older beds in a definite stratigraphic sequence or vertical succession. Thus was born the science of geological stratigraphy. There are exceptions, however, to this stratigraphic rule of *superimposition*. One common in mountain belts is provided by recumbently-folded strata (Fig. 4.5B). Older beds are seen in the lower limbs of such folds to overlie younger, and special criteria have to be sought in the field in order to establish the true 'way-up' of the strata. Another exception occurs in lava fields, where empty lava tubes, which once funneled magma away from the crater of a volcano, became filled up by younger, washed-in sediment. Deposits accumulated in rock-shelters and caves afford a further instance.

An archaeologist recording the 'stratigraphy' of a site does so using the same principles as the geologist: younger normally overlies older. A common exception to the rule, exactly like the caves and lava tubes above, is to found in *building stratigraphy*. A Norman window in a Norman church wall may have been replaced by a fourteenth-century Decorated one. Although not exactly exceptions to the principle of superimposition, there are certainly circumstances that can lead to confusion, for example, where an archaeological sediment contains artefacts from an earlier time that have been redeposited, either by human agency or natural processes. A careful search for truly contemporaneous items is then necessary. A geological parallel is afforded by the reworking of Carboniferous spores – microscopic and strongly resistant to decay – into Cretaceous sandstones millions of years younger. The only difference that can be found between geological and archaeological stratigraphy is that geological time is vastly deeper than the archaeological record.

One of William Smith's most profound discoveries at the turn of the eighteenth century was that sedimentary strata, especially those of marine origin, could be *correlated* from place to place, that is, identified as of the same age, by means of their contained assemblages of fossils. This was found to be true even where the rocks themselves were changing lithologically along the line of their outcrop or cropped out in unconnected places. A corollary of these findings was that, in sequences that lacked significant depositional breaks, the character of the fossil assemblages often seemed to change gradually as younger beds succeeded older.

A common practice amongst archaeologists is to demonstrate the geographical range of a culture by mapping out the sites attributable to that culture: where the artefact assemblages are found to be the same, then the sites can be referred to the same culture. In the archaeological as in the geological case, the process of referral involves reference, normally implicit once the field has been opened, to some type-site where the culture was first recognized or was especially well described. As with fossil assemblages, artefacts such as pottery often gradually undergo subtle changes of style, and so can be used for relative dating (seriation), as W.M. Flinders Petrie (1853–1942) discovered in Egypt.

Systematic stratigraphy

The geological maps discussed in Ch. 4 each include a column of strata in which the different beds, arranged in upward order of decreasing age, are named as *mappable lithological units* (Formations, Members, Beds), and are assigned to particular portions of geological time. In other words, the stratigraphy of the rocks represented on the maps is systematized. The intervals of time recognized by geologists are *Eras*, divided into *Periods*, in turn split into *Epochs*. The geological *Systems* are the bodies of rock formed in particular periods. The Devonian System, for example, comprises all the rocks formed during the time-interval called the Devonian Period, and takes its name from the County of Devon, where these rocks were first described.

Table 5.1 is a brief summary of the systematic stratigraphy of Britain and Ireland, together with the main geological events and conditions that affected the islands. The earliest rocks, referred to as the *Riphean Era* and as the *Archaean* and *Proterozoic Periods*, are chiefly found in the Scottish Highlands and islands. The oldest of these are intensely metamorphosed sediments and igneous rocks dated to between 3.2 and 1.7 billion years old, that is, some 1.6 billion years after the Solar System was formed. Simple marine fossils began to appear in the Proterozoic.

The following short interval of time, known as the *Precambrian* or *Vendian*, sees the emergence of algae and soft-bodied invertebrates known only as impressions. The shallow-marine rocks of this age are represented in Britain by the Charnian sediments of Charnwood Forest in the East Midlands, in which a grammar-school boy discovered the characteristic fossils.

The soft-bodied, Vendian organisms were quickly followed in the *Cambrian Period* by a huge variety of highly diverse organisms with, for the first time, readily preserved shells and hard parts. Vertebrates began to appear in *Ordovician* times. Like the *Ordovician* and *Silurian Periods* that followed, the Cambrian Period is named after rocks outcropping in the territory of a Welsh tribe encountered by the Romans. It was in north Wales, settled by this tribe, that Cambrian rocks were first recognized; the Ordovices and Silures lived further south but also gave their names to important sections of geological time. The Silures, for example, had a tribal centre at the small, walled town of Caerwent in southeast Wales (Ch. 9, Case 2). The Cambrian, Ordovician and Silurian Periods together comprise the *Lower Palaeozoic Era*. The rocks of this Era, found in Wales, the Welsh Borders. Cumbria, the Southern Uplands of Scotland and parts of Ireland, are of mixed marine and volcanic origin. Metamorphosed Lower Palaeozoic and later Proterozoic sediments underlie the Scottish Highlands.

The *Upper Palaeozoic Era* is made up of the *Devonian, Carboniferous, Permian*, and *Triassic Periods*. The leafless land plants that had begun to appear in the late Silurian evolved rapidly into trees with naked seeds, for example, conifers and cycads. Insects together with fish and amphibians emerged in the Devonian, represented in Britain and Ireland chiefly by the non-marine (fluvial, lacustrine) *Old Red Sandstone* facies. Abrupt climate change reintroduced marine conditions during the early Carboniferous, followed by the enormous, humid coal swamps and deltas of the later Carboniferous, inhabited by large insects and reptiles. The Millstone Grit and Coal Measures formed in these deltas and swamps have afforded sandstones for building and paving towns and cities throughout South Wales, the Pennines and the Scottish Midlands. Non-marine Devonian and marine Carboniferous rocks dominate the geology of Ireland. Desert conditions appeared in the Permian and, later, the Triassic.

The *Mesozoic Era* comprises the *Triassic, Jurassic* and *Cretaceous Periods*. As Williams Smith's map shows (Fig. 4.1), rocks of these dates combine to form a great sash-like outcrop that can be traced across England from the Channel coast in Dorset and the Strait of Dover to the shores of Yorkshire on the North Sea. The great escarpment and dip-slope that includes the Cotswolds is formed by creamy-yellow, shelly and oolitic limestones of Jurassic age, much sought after in Roman and later times as a building and architectural stone. The Cretaceous Chalk Group gives rise to perhaps even more extensive escarpments in England. Flowering plants are numbered amongst the characteristic fossils of the Mesozoic, together with corals resembling modern forms, dinosaurs, birds, bony fish, and primitive vertebrates. Mesozoic sediments are few and far between in Scotland and confined to the eastern and western coastal margins.

What typifies the *Cenozoic Era*, leading up to our own times, is the presence of organisms that look modern, including many genera and even some species that are extant today. The *Tertiary Period* is divided into five Epochs. The earlier of these are well-represented by the shallow-marine sediments found in the Thames and Hampshire Basins and by the lava fields and igneous complexes of Northern

Era	Period	Epoch	Date (Ma BP)	Geological events
Cenozoioc	Quaternary	Holocene	0.012	Glaciers recede. Sea-level rises. Climate equable, mild and maritime
		Pleistocene	2.6	Repeated climate fluctuations causing glaciers to form and recede, permafrost to form and melt, and sea-level to fall and rise dramatically
	Tertiary	Pliocene	5.2	Continents and oceans essentially in present configuration. Ice caps form
		Miocene	28.3	Seas recede further. Mediterranean Sea evaporates. Asia and Europe fuse
		Oligocene	35.4	Seas recede. Alpine and Himalayan mountains form as Africa drives north. Some related folding in southern Britain
		Eocene	56.5	Greenland and Australia separate from Laurasian continent. Climate subtropical/tropical. Shallow-marine sedimentation in southern England
		Paleocene	65	Volcanicity in Northwest Scotland. Flood basalts in northern Ireland and the Western Isles
Mesozoic	Cretaceous		97.5	Chalk thickly deposited in extensive shelf seas. Estuarine-lacustrine then shallow-tidal sedimentation in southern England. South America and Africa separate
	Jurassic		144	Seas advance. Climate subtropical. Coral reefs and oolite shoals at times but seas mainly sandy or muddy. Deltas with coal swamps in northeast England and eastern and western Scotland. North America and Africa separate
	Triassic		213 248	Hot climate becoming moister. Desert conditions. Alluvial fans, dunes, evaporating gulfs and lakes (halite). Pangean supercntinent breaks up into Laurasia (north) and Gondwana (south)
Upper Palaeozoic	Permian		286	Hot climate with evaporation of shallow seas (dolomite, polyhalite). Dunes. Southern hemisphere glaciation. Folding in southern Br. Isles
	Carboniferous		360	Enormous alluvial plains, deltas and coal swamps follow deposition of limestones in extensive, shallow, spreading seas. Granite intrusions in Southwest England and the Midlands
	Devonian		408	Mountain-building caused by continental collision preceded and followed by alluvial sedimentation. Fluctuating lake in northeast Scotland and N. Isles Widespread volcanicity in southern Scotland

	Silurian	438	Marine sedimentation almost everywhere with deepest waters in northeast Ireland, southern Scotland, Cumbria and Wales. Some volcanicity
Lower Palaeozoic	Ordovician	505	Marine sedimentation almost everywhere with deepest waters in northeast and southeast Ireland, southern Scotland, Cumbria and Wales. Numerous volcanic centres in Wales and Cumbria. Folding and metamorphism in Scottish Grampian Highlands. Europe and North America move together
	Cambrian	590	Extensive shallow seas in the Scottish Highlands and English Midlands. Deep water in most of Wales. Local volcanicity
Precambrian	Vendian	650	Climate warm. Shallow seas advance and retreat
Riphean	Proterozoic	2500	Shallow shelf seas with carbonate deposition. Glaciation. Fluvial redbeds deposited in Northwest Scotland. Glaciation. Intense deformation and metamorphism
	Archean	4600	Intense bombardment and cratering. Earth's crust and oceans formed. Banded Ironstone formations deposited

Table 5.1 Summary of the stratigraphy of Britain and Ireland and the chief geological conditions and events

Ireland and the Western Isles of Scotland. During the *Quaternary Period*, divided between the *Pleistocene* and the *Holocene* – some stratigraphers would now unofficially add the *Anthropocene* as the latest epoch – there were profound fluctuations of climate that led to repeated glaciation and sea-level change. Geology and archaeology merge in the Pleistocene and Holocene. Archaeologists have created for Britain and Ireland a parallel systematic stratigraphy of the Quaternary: Palaeolithic, Mesolithic, Neolithic, Bronze Age, Iron Age (including Roman), Medieval, Early Modern and Modern Periods. Geologically, human activities in Britain date from at least 700,000 years ago in the Pleistocene, at localities such as Pakefield and Happisburgh in East Anglia, but occupation has not been continuous.

Geological and archaeological dating

Superimposition, assemblages of fossils or artefacts, and seriation only provide geologists and archaeologists with *relative* dates. It is highly desirable, however, to have *absolute* dates for geological and archaeological materials and events. Obtained using a wide range of science-based methods, several are mentioned below.

The most important and reliable are *radiometric* and depend on the behavior of elements with radioactive isotopes.

The atom is the smallest part of an element that can take part in a chemical reaction. It is modelled as a nucleus, of neutrons and protons, surrounded by electrons orbiting in discrete shells. The number of protons in the nucleus describes an element in terms if its *atomic number.* The number of neutrons, however, determines its *atomic mass* or *weight.* Many elements consist of *isotopes* that, although chemically identical, differ in the number of neutrons present and therefore slightly in their atomic weight. Isotopes are of two kinds. Those described as *stable* do not undergo spontaneous radioactive decay, but nonetheless are valuable in some cases as indicators of diet, nationality, and palaeotemperature. *Unstable isotopes*, or *radionuclides*, disintegrate spontaneously into a daughter isotope, with the simultaneous emission of an energetic particle (alpha, beta) or electromagnetic radiation (gamma). Radioactive decay takes place at a fixed, constant, measurable rate distinctive of the element, and is measured by the *half-life.* The half-lives of the radioisotopes chosen for dating by geologists and archaeologists vary between a few years and a few billion years, always a matter of horses for courses when selecting a method.

The first person to successfully use radiometric dating in geology was (Sir) Arthur Homes (1890–1965). Raised in the northeast of England, and trained as a physicist and geologist in London, Holmes in 1911 succeeded in dating a Norwegian rock to the Devonian period, by exploiting the fact that the element *uranium* present in its minerals has radioactive isotopes that decay into lead. The extreme depth of geological time, postulated by such as Hutton, and the likely age of the Earth, was thereby conclusively demonstrated. Today, in addition to uranium and its decay series, geologists and archaeologists have at their disposal means of radiometric dating using isotopes of potassium and carbon.

Geologists have had great success in using radioactive uranium to date the Earth's rocks, including the oldest known, for uranium-238 (^{238}U) and uranium-234 (^{234}U) have half-lives of respectively 4.5 billion years and 245,000 million years. Using a spectroscope, the method involves measuring the isotopes present in preferably igneous rocks such as lavas found in stratigraphically well-defined positions. Two of the daughter isotopes in the uranium decay series are thorium-230 (^{230}Th) and protactinium-231 (^{231}Th). With half-lives of just 75,400 years and 34,300 years respectively, these radionuclides have archaeological applications, for example, the dating of Pleistocene and Holocene cave speleothems, bones, teeth, molluscs, and corals.

The alkali metal *potassium* has a radioactive isotope ^{40}K with a half-life of 1250 million years. Part of the decay leads to the gas argon ^{40}Ar, by the emission of gamma rays, and another part to the alkaline earth metal calcium ^{40}Ca, by the creation of a beta-particle. Dating using potassium-argon has been successfully applied by geologists to the younger parts of the geological record and by archaeologists to lavas or volcanic ashes that preserve artefacts.

Dating by means of *radiocarbon* is cheap, reliable, and now routine in archaeology where organic materials such as peat, wood, leather, and bone are encountered in Holocene and latest Pleistocene contexts. Carbon has a stable isotope ^{12}C and a radioactive one ^{14}C that changes into ^{14}N with the emission of one beta-particle from each atom (half-life 5730 years). Radiocarbon is formed in the upper atmosphere when cosmic rays interact with the nitrogen atoms present. From the atmosphere radiocarbon is transferred by photosynthesis into plants, and indirectly into terrestrial animals, and also is exchanged with carbon dioxide in the oceans to enter the shells and skeletons of marine organisms.

The above radiometric techniques available to archaeologists are the most direct, because they do not require calibration. Some level of independent calibration or reference to context is called for by many other methods of dating. A group of counting methods based on annual effects exploit *varves* in lake deposits, *snow layers* in ice cores, and *tree rings*. These can yield exceptionally precise results when combined with radiocarbon dating and statistical manipulation. The Earth's magnetic field is neither constant in either strength or polarity, changes that have been exploited as a dating tool. Other dating methods that have been developed are thermoluminscence (burnt/fired materials) and optically-stimulated thermoluminescence (unheated sediments) dating, the hydration of obsidian, the rehydration of brick and pottery, and the gradual degradation (racemization) of amino-acids in organic matter (mollusc shells). The search for further methods goes on.

Further Reading

Aitken, M.J. 1990. *Science-based dating in archaeology*. London, Longman
Anderton, R., Leeder, M.R. & Sellwood, B.W. 1979. *A dynamic stratigraphy of the British Isles*. London, Allen & Unwin
Lewis, C. 2000. *The Dating Game. One Man's Search for the Age of the Earth*. Cambridge, CUP
Toghill, P. 2000. *The geology of Britain*. Marlborough, Crowood Press.

6. Geology and Landscape

What is landscape?

In its broadest sense, and as a spatial concept, landscape is the total appearance of the land in a place, district or region: the rocks, soils and minerals, the shape of the land and the scale of its features, its vegetation and land-use, the pattern and kinds of settlement, its industrial elements, and even the general appearance of the sky. In principle, there are two end-member kinds in Britain and Ireland: landscapes of wholly natural origin and others that are wholly cultural, encompassing the creations and modifications due to humankind. Landscapes are affected by many, interacting factors but at the root lie geology and geological processes (Fig. 6.1).

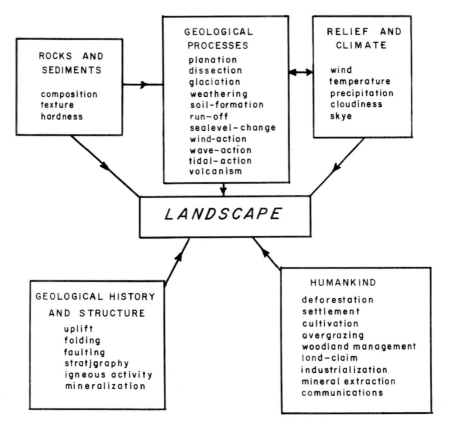

Fig. 6.1 The chief influences on landscape

Geological history

With some rocks 3.2 billion years old, the geological history of Britain and Ireland is neither simple nor straightforward (see Table 5.1). In essence, however, it revolves around the history of *four oceans*. The earliest, existing during the Lower Palaeozoic Era, had on one side what are now Wales, Cumbria and much of Scotland, but with northwest Scotland on the opposite shore. This ocean had closed by early Devonian times, with the fusing of the two margins, accompanied by metamorphism and the intense folding of the thick marine sediments on northeast-southwest lines. A new ocean then came into being which had most of Britain and Ireland situated on an extensive land-mass to the north, on which a range of shallow-marine, fluvial, lacustrine, and deltaic sediments were deposited. Only in Southwest England are the deep-marine Devonian and Carboniferous sediments of this new ocean encountered at outcrop. In late Carboniferous and Permian times the sediments deposited in this ocean became intensely folded on east-west lines as this ocean closed to form a new continent called Pangea. A third ocean, called Tethys, appeared within the vast extent of Pangea. Its sediments, when folded, metamorphosed and thrust, formed the mountain chains of the Alps. In southern Britain, which lay nearest but far to the north, only faint ripples of disturbance due to this event were felt. Much more important was the opening and widening of the Atlantic Ocean during the early Cenozoioc, to the accompaniment of great volcanism in Ireland and western Scotland. The spreading of the Atlantic Ocean continues today, but more passively.

The crust in Britain and Ireland is therefore crossed by two major fold-belts, the northeast-southwest *Caledonides* in Scotland, Cumbria, Wales and eastern Ireland, and the *Variscides* ranging west-east from southern Ireland, through southern Wales and Southwest England and, concealed at depth, across southern England to reappear in the near-continent. The creation of each fold-belt was accompanied by the widespread intrusion of granites. The accompanying mineralization underpins many of today's metallurgical landscapes (Ch. 12).

Structure, rocks and sediments

Taking a very broad approach, Britain and Ireland may be divided geologically into two parts so far as landscape and many resources are concerned: an 'Old' and a 'Young' realm (Fig. 6.2).

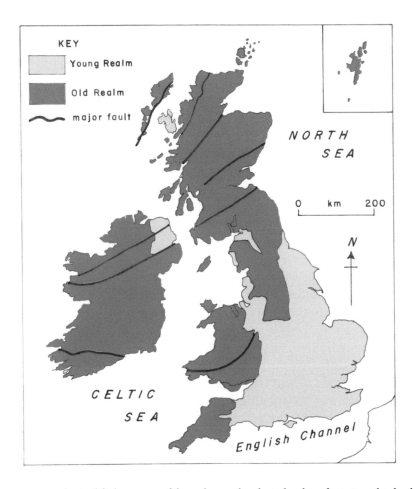

Fig. 6.2 Greatly simplified summary of the geology and geological realms of Britain and Ireland

Old Britain and Ireland are formed of pre-Cambrian, Lower Palaeozoic and Upper Palaeozoic rocks, up to and including the Carboniferous. The older of these rocks were intensely folded, thrust, metamorphosed, and intruded by granites when the Caledonides were formed. The later ones found in southern Britain and Ireland were similarly influenced when the Variscides were created, again with the intrusion of granites. The Devonian and Carboniferous strata outcropping further north were less affected, but nonetheless became folded and faulted.   Because of their geological history, the hard, slow-weathering rocks of Old Britain typically give rise to uplands with thin, poor, acidic soils to which forestry, moorland, and grassland and livestock-farming are best-adapted. As a sort of compensation, the resources available in this realm are much stone and slate for building and a wide range of important metalliferous ores.

The Young realm is formed chiefly of Permian, Triaassic, Mesozoic and Cenozoic sediments. In the main, these deposits are soft and weather readily, creating rich, well-drained, neutral soils. With the exception of regionally extensive scarps formed by Middle Jurassic limestones and the Chalk Group, they give rise to low lying landscapes of gentle relief. Mixed arable farming and livestock husbandry is practiced in western parts of the realm. The cultivation of cereal and root crops dominates in the east and much fruit is grown locally; livestock here are kept chiefly on outcrops of river alluvium. The only significant metalliferous deposits encountered are those of iron. Stone for building is limited in its availability and mainly restricted to Permian and Jurassic limestones and flint. On the other hand, excellent potting and brick-making clay are found in abundance.

Geological processes

Geological processes that operate at many different time and spatial scales affect landscape. They may be divided between *internal processes*, operating within the Earth's crust, and *external processes*, directly influencing the ground surface. The two are interconnected. For example, crustal uplift promotes the development of relief that in turn enhances weathering, run-off and sediment yield.

Crustal uplift on a geological time-scale has significantly influenced the landscape of Britain and Ireland. The Boxgrove Middle Pleistocene hominid site on the southern face of the South Downs stands on an extensive shoreline now *c.* 40 m above present sea level (see Ch. 8), implying an average rate of *uplift* approaching 0.1 mm per year. The Downs as a whole, and the chalklands of southern England and northeast and central France, are capped by a distinct, gently undulating plateau at a maximum elevation of the order of 200 m. A similar feature, but at lower altitude, can be traced throughout southwest Wales to the west and south of the Preseli Mountains. These features are the product of *planation*, possibly marine, many millions of years ago, but which have been elevated as the result of subsequent, regional but not uniform crustal uplift. Planation, as a general surface process, sees the slow retreat of a cliff or escarpment due to episodic collapse followed by the degradation and removal of the fallen material. One result of planation followed by uplift is that drainage systems become incised, as witness Milford Haven and its many branches in southwest Wales and the Helford, Fal, Tamar-Tavy, Dart and Exe estuaries in Southwest England. Several of these well-sheltered waterways have become important port-settlements. Flights of river terraces may develop during river incision, as witness the gravels of the Thames Valley, with their rich yields of transposed Palaeolithic flintwork.

Glaciation and cold-climate (periglacial) conditions over the last few million years have profoundly affected the regional landscape of Britain and Ireland. *Glacial erosion* has shaped all but the southernmost uplands, creating the deep, steep-sided U-shaped valleys, many adorned with lakes or lochs, that typify southwest Ireland, Wales, Cumbria and the Scottish Highlands. High at the heads of these valleys, especially on north- and east-facing slopes, are found *cirques*. These are large, amphitheatre-like landforms that trapped snow and fed the glaciers further downslope. The minor landscape features of wholly or partly erosional origin to be found in these valleys are glacially smoothed and striated rock surface, whalebacks, *roches moutonées*, and crag-and-tail. *Glacial deposition* is in many ways less obvious in its expression (see also Ch. 3). Most of lowland England is blanketed by *subglacial till*, accompanied by other glacier-related sediments, sands and gravels. In midland and northern Ireland, the Midland Valley of Scotland, and the Cumbrian lowlands, the surface of the till has been moulded by moving ice into fields of low, cigar-shaped, parallel hills on a kilometre scale called *drumlins*. As the result of rising sea-level many are now found as islands, those of Strangford Lough, Northern Ireland, are especially impressive. *Surpraglacial tills* are much more restricted in occurrence. Typically, they underlie groups of mounds or arcuate ridges within or at the lower ends of glaciated valleys. Good examples in Cumbria can be seen at Rossthwaite and to the north of Ambleside. Kettle holes, now ponds or meres where ice masses stagnated, and trains of huge boulders – glacial erratics – are common associates of moraine.

At outcrop crustal rocks experience conditions of temperature, pressure, moisture and oxygenation different in varying degrees from those under which they arose. The rocks are consequently out of chemical and physical equilibrium with the surface environment, and therefore open to weathering and soil-formation, the slow transformation of the rock into discrete mineral particles, new substances, and dissolved materials by interconnected physical, chemical and biological agencies. The resulting material is *soil*. Although commonly showing a vertical sequence of contrasting layers or *horizons*, soils are not sediments, but being formed in place are in their origin more akin to metamorphic rocks. Their character is strongly influenced by relief, parent material, and microclimate.

The soils of Britain and Ireland that formed during the Holocene are of many kinds, but may be illustrated by considering two important, contrasting groups. *Podzols* are typical of cool, humid Ireland and the west of Britain. They are the products of strong leaching by rainwater and the downward movement within the soil profile – *translocation* – of suspended colloidal and dissolved substances. The topmost horizon in a podzol is formed of plant litter decaying to yield humic acids. As these acids leach downward, they carry with them colloidal organic matter and break down clay minerals to yield iron and aluminium compounds, and dissolved alkali and alkaline earth metals. A grey to purplish-grey bleached horizon results. In the next horizon down, coloured a strong orange-brown, the iron and aluminium compounds become precipitated, and there may also be some accumulation of colloidal organic matter above the coloured layer. On a steep enough slope, the soils' constituents are transferred partly downslope as well as purely downward, giving rise to a lateral sequence of slightly contrasted soils known as a *catena*. By contrast, on the calcareous substrates provided by the Carboniferous Limestone, the Middle Jurassic limestones, and the Chalk Group much simpler soils known as *rendzinas* develop. These have only one horizon, grey to reddish-brown in colour, with moderate organic matter, and fragments of the parent rock. The base is sharp, often in the form of steep pinnacles and deep pits, clearly the result of the solution of the carbonate rock, which may in detail become pock-marked and riddled with tunnels.

Soils on sloping ground are subject to very slow downslope movement under gravity, known as *soil-creep*. Under cool conditions the rate of downslope movement is greatly enhanced by repeated freeze-thaw, leading in cold regions to *solifluction*.

Climate

Climate may be regarded as the general character over a protracted period of the weather in a locality, district or region. The climate in some localities, for example, a sunlit, south-facing slope, or a sheltered valley, is often distinguished as a *microclimate*, something exploited by farmers when selecting crops to grow. The weather system itself is highly variable, changing daily and seasonally, and also on decadal-century time-scales. The British and Irish climate, best summed up as temperate-maritime, is determined by *wind*, *temperature*, *precipitation* and at least the *cloudiness* of the atmosphere

The islands receive winds from all directions but, lying within the zonal *Westerlies*, experience the most frequent and powerful winds from the northwest, west, and southwest. Storms cross the Atlantic from these directions. Because of the proximity of the open sea, the coasts are the windiest places, but those in the west and north are the stormiest. The wind mostly affects coastal landscapes, promoting coastal erosion and sand-blow, and limiting tree-growth. Salt spray drifted inland influences what can be cultivated in the coastal zone.

The *temperature* in Britain and Ireland is influenced by the Gulf Stream in the west and north, and proximity to mainland Europe in the southeast. Mean-annual temperatures during winter decline by a few degrees Centigrade northeastward from southwest Ireland, southwest Wales, and Southwest England to eastern England and Scotland. In summer the fall is from southeast to northwest, again by a matter of a few degrees Centigrade. Temperature declines with increasing altitude at all times of year, promoting physical over chemical rock-weathering. These spatial patterns, together with precipitation, strongly influence land-use, favouring arable-farming over animal husbandry in the drier and warmer regions.

Especially strong spatial gradients are shown by *precipitation* as rain or snow. The mean-annual value is as little as 40-60 cm in eastern and midland England and in parts of eastern Scotland, rising to three to four times that amount in western Britain and Ireland. Like temperature, precipitation is strongly influenced by altitude. Extreme values of 2000 cm or more are experienced in the snowy uplands and mountains of Scotland, Cumbria, the north Pennines, Wales, Southwest England, and the extreme west of Ireland. High precipitation encourages high run-off and sediment yields, soil-leaching and acidification, peat growth, and soil waterlogging. Forestry and livestock-farming are favoured over the growing of crops.

Cloudiness limits the amount and frequency of sunlight received at the Earth's surface. It has an indirect affect on land-use, tending to be least at the coast and, regionally, greatest in the west and north. Like precipitation, cloudiness increases with altitude.

Humankind

Human activities have profoundly influenced the landscapes of Britain and Ireland. Indeed, it is arguable that there are no truly natural landscapes left for scrutiny, and that a cultural element is perceptible in all landscapes, even if just tracks worn by ramblers up to mountain-top cairns. Woodland management, with *deforestation* as its extreme, has been practised since the Mesolithic. The complete removal of woodland from an area changes the local climate and flora and, if followed by arable farming, leads to soil loss and increased stream flashiness, run-off and sediment yields. The introduction of livestock-farming, especially if there is over-grazing, also increases run-off and, altering the local flora, can lead to a severe loss of biodiversity. Many moorlands are now managed for game.

As discussed in Ch.8, many tidal marshes have been subject since Roman times to *land-claim*, that is, made suitable for arable farming and settlement by the construction of seabanks and drains. The resulting enhancement of economic potential has been at the expense of the loss, in these liminal environments, of valuable, essentially marine resources, and the cost of constant drainage and maintenance of the defences.

The process of *industrialization* can be traced from early times but had most effect as a landscape-changer from the eighteenth century onwards, with the opening of quarries and mines, and the building of manufactories, ports, canals, turnpike roads, and railways. Large parts of the country are clearly industrial landscapes, if partly or wholly relict because the industries that marked them have closed down and the settlements that depended on them have turned, not always successfully, to other sources of income. Examples are the Swansea and Bath copper-brass foundries, 'metal-bashing' in Birmingham and the 'Black Country', the cotton and wool towns of Lancashire and South Yorkshire, and the shipyards of Belfast, Clyde, and Tyne. These and many other regions present a wealth of industrial archaeology ripe for study.

Great importance attaches to *extractive industries*, such as coal-mining in South Wales and Durham, and salt-production at Droitwich and in Cheshire. Some kinds of building stone have been extensively worked since medieval times, for example, near Bath and Peterborough (Ch. 9). The mining of metalliferous ores was an important activity, especially in 'Old' Britain and Ireland (Ch. 12). Cornwall produced tin from before Roman times. The Cumbrian massif, for example, abounded in active mines and quarries (copper, lead, zinc, iron, graphite) during the eighteenth and early nineteenth centuries, while at the same time being the inspiration of Romantic poets, and the locus of the Sublime and the Picturesque.

Further Reading

Audouy, M. & Chapman, A. 2009. *Raunds: the origin and growth of a medieval village.* Oxford, Oxbow Books

Barron, A.J.M., Morigi, A.N. & Reeves, H.J. 2006. *Geology of the Wellingborough district: a brief explanation of the geological map sheet 186.* Keyworth, British Geological Survey

Bracegirdle, B. 1973. *The archaeology of the industrial revolution.* London, Heinemann

Bridges, E. M. 1978. *World soils.* Cambridge, Cambridge University Press

Friend, P.F. 2008. *Southern England. Looking at natural landscapes.* London, Harper-Collins

Friend, P.F. 2012. *Scotland. Looking at natural landscapes.* London, Harper-Collins

Garlick, T. 1985. *Hardknott Castle Roman fort.* Lancaster, Dalesman Books

Janson-Smith, D., Cressey, G, & Fleet, A, 2008. *Earth's restless surface.* London, The Natural History Museum

Leeder, M & Lawler, J. 2016. *GeoBritannia: Geological Landscapes and the British People.* Edinburgh, Dunedin

Smith, A. 2008. *The Ice Age in the Lake District.* Keswick, Riggside Publications

Waterhouse, J. 2004. *The stone circles of Cumbria.* Chichester, Phillimore

Case 1: Pleistocene glaciation of the Cumbrian massif

Geologically, the Cumbrian massif is a highly complex dome of Lower Palaeozoic rocks, metamorphosed to slate grade, ringed by gently outward-dipping Carboniferous and Permo-Triassic beds. The Lower Palaeozoic rocks have a complex, broadly anticlinal structure that strikes northeast-southwest. The present doming is likely to be Cenozoic.

The land rises from the surrounding lowlands to almost 1000 m in much of the massif. Three landscape zones are recognizable. Rounded fells (e.g. Skiddaw, Caldbeck) typify the slates of the northwest. Striking across the middle of the massif are the rugged, craggy mountains and fells, such as the Langdale Pikes, formed by the Borrowdale Volcanic Series. The southeastern fells are composed of interbedded sandstones, slates and some limestones that create mixed rounded and craggy landforms. Life was never easy in these wet uplands, but sites of most dates have been recognized. There are, for example, more than 60 Neolithic-Bronze Age stone circles and related monuments known from within and on the outskirts of the fells. Castlerigg with its lengthy sitelines, prominently situated near Keswick, is perhaps the best known. The region has been intensively quarried and mined for metals and graphite, and is renowned for its fine mineral specimens.

During glaciation ice streamed radially outward in all direction from the centre of the massif, gouging deep, U-shaped valleys like the spokes of a wheel, many of which now hold deep lakes (Fig. 6.3). High up at the heads of the valleys, chiefly on slopes facing north or east, are cirques, often with tarns. These features are most conspicuous on the outcrop of the strong Borrowdale Volcanic Series rocks. On leaving the massif the ice flowed northwestward along the Vale of Eden to the east, shaping a drumlin landscape, before circling west and southwest through the Cumberland Lowlands to the north, where another, even larger drumlin field was formed. Ice-flow directions are revealed by small-scale features, such as striae, whalebacks and *roches moutonée*, and by the elongation of the drumlins. The patterns formed by latter are clearest in satellite imagery.

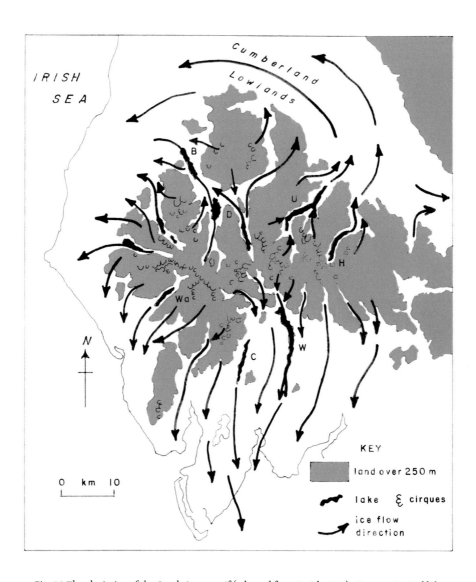

Fig. 6.3 The glaciation of the Cumbrian massif (adapted from Smith 2008). Key to principal lakes:
B – Bassenthwaite Lake; C – Coniston; D – Derwewntwater; H – Haweswater; L – Loweswate;
U – Ullswater; Wa – Wastwater; W – Windermere

Case 2: Hardknott Castle Roman fort, Cumbria

In Hadrianic times the Roman army built three forts in the southwest Cumbrian massif: at Ravenglass with a port on the coast, at Ambleside at the head of Lake Windermere in the far northeast, and in the western foothills of the Hardknott Pass linking these two. This fort (*Mediobogdum*) was sited with a very sharp eye to the strategic advantages offered by the local terrain (Fig. 6.4). It lies on a gently shelving, wedge-shaped spur of ground at a height of 250-300 m that projects southwestward from the Border End (522 m) and Hard Knott (549 m) fells. To the northwest is the steep-sided, alluviated, glacial valley of the R Esk flowing to Ravenglass on the coast, extensive views toward which are available from the fort. Steep slopes and precipitous crags defend the fort from attack from this valley. To the southeast the fort commands a view over the deep V-shaped valley of Hardknott Gill and the first steep slopes of the winding route over the Pass. Attack from the northeast was virtually impossible. The fort itself, built of local rocks, is very well preserved, of a square plan, with rounded corners, internal corner-tower, and outbuildings. Some 75 m to the northeast is a cleared and levelled area identified as a parade-exercise ground.

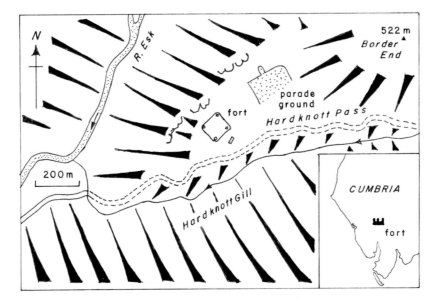

Fig. 6.4 The setitng of Hardknott Castle Roman fort in southwest Cumbria

Case 3: The Nene valley, Wellingborough, Northamptonshire

The R. Nene rises to the west of Northampton and winds its way northeastwards across the plains of the southeast Midlands to reach the sea in The Wash of East Anglia. The town of Wellingborough lies in the middle reaches of the Nene, at the confluence with a tributary, the Ise, entering from the north. The rivers occupy broad, alluviated valleys incised by 40 m or so into an extensive, level plain of little relief.

The geology of the area has profoundly influenced settlement and land-use within it (Fig. 6.4). The valley slopes expose a sequence of Jurassic rocks from the Upper Lias in the west to the Oxford Clay in the east with a very gentle eastward regional dip. The dissected interfluves are cloaked by a thick blanket of Pleistocene till, grey with Jurassic and Cretaceous clasts below becoming chalky upward. Locally there are Pleistocene sands and gravels.

Weirs (W), perhaps for fishing, and mills (M) lie scattered along the rivers, which in the case of the Nene had been locally straightened to improve navigation and flood control. Their adjoining floodplains are underlain by locally peaty alluvium that offers excellent grazing. Settlement, however, took place mainly on the valley sides and shoulders, where strings of medieval villages of Saxon origin, including much-excavated Raunds and Higham Ferrers, can be found. These slopes have well-drained, easily-worked, neutral soils formed on a complex sequence of Middle Jurassic sands, silts, clays, ironstones, and limestones. The digging of iron ores and iron smelting have been practised extensively on these slopes since Roman times. In sharp contrast, the Pleistocene till on the interfluves affords heavy, ill-drained soils difficult to cultivate by traditional methods. Occasional hamlets and thinly dispersed cottages and farmsteads are the only settlements found here.

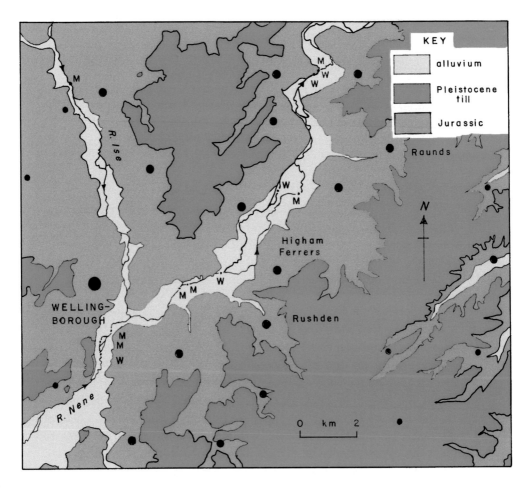

Fig. 6.5 Geology and ssettlement in the Nene Valley, Wellingborough.
Adapted from British Geological Survey 1:50,000 Sheet 186

Case 4: Land-claim in the Nene estuary, Fenland, eastern England

By the time of King William I's Domesday Survey (1086) the seabank protecting the silt Fenland defined a funnel-shaped estuary at the mouth of the R. Nene, with the medieval-early modern port of Wisbech at its head (Fig 6.6). At Leverington, just northwest of the town, a well-preserved section of this bold, but now redundant, seabank can still be seen. Strings of compact, nucleated villages with Saxon origins, such as Tydd St. Giles, Long Sutton, West Walton, Walpole St. Peter and Terrington St. Clements, line this coastline to the north and northeast of Wisbech beneath open East Anglian skies. Subsequent aggressive land-claim of tidal marsh and mudflat rapidly closed up the estuary and forced the mouth of the Nene increasingly northward. The gain averaged of the order of 0.25-0.50 km^2 per year as parcels of marsh, typically 1-3 km^2 in extent, were embanked in sequence by landowners in the area. At the same time, the Nene was straightened and a loop of the R. Great Ouse was cut off by the Eau Bank Cut of 1821. A quite different and much less dense settlement pattern compared to the early medieval one obtains in these highly productive areas. Typically, the sparse farms and cottages lie out of the reach of floods along tracks and roads mounted on the seabanks, forming thin, linear settlements. Sutton Bridge on the Nene is the most substantial. The adjacent estuary of the Great Ouse and the medieval port of King's Lynn were similarly affected, although later and on a lesser scale. Land-claim ceased in the late twentieth century, as it was considered to be damaging the regime of The Wash.

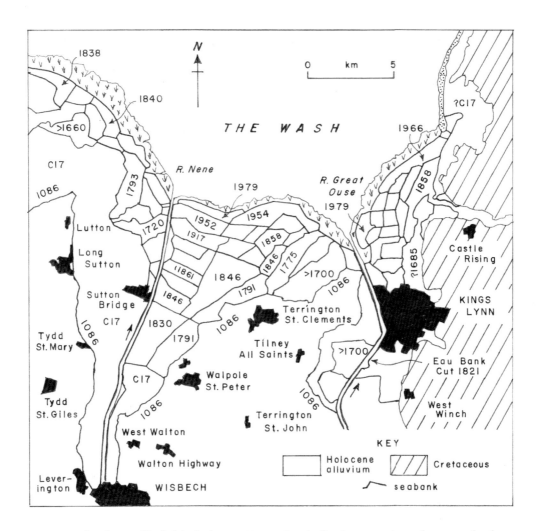

Fig. 6.6 The advance of land-claim in the Nene Estuary, East Anglia. Representative enclosures are dated

7. Rivers and Water Management

Rivers and their character

Rivers and streams are turbulent, channelized flows of water across the Earth's surface, flowing from high to low ground. Gravity 'pulls' the water down each sloping channel against the frictional resistance offered by the bed and banks, potential energy being in the process exchanged for kinetic energy. The function of a river is to drain rainfall and melt-water from its catchment and to carry toward the ocean the particulate and dissolved products of the weathering of the rocks outcropping within its catchment. The river thus completes the *water cycle* and initiates the *sediment cycle*, two crucial, continuous, connected processes of the outer Earth. Some of the debris that enters rivers lodges in their valleys, deltas, and estuaries, providing would-be settlers with fertile landscapes and sites suitable for agriculture, fishing, industry, and trade, albeit subject to occasional flooding.

Britain and Ireland have many rivers but they are mere trickles by global standards. Comparing them, they may be ranked in terms of *length*, *catchment area*, and *mean annual water discharge*. The Thames is first in length at 239 km and also first in catchment area at 9950 km^2. In terms of mean annual discharge, the Tay, rising in the Scottish Highlands, is ranked first with a mean annual discharge of 152 m^{3}s^{-1}, more than twice the value for the Thames and Severn, and almost twice that of the Trent. The daily discharge of British and Irish rivers is markedly unsteady and varies on a seasonal cycle. The least values are typically seen in summer, when precipitation is low and evapotranspiration high. Peak values mark the cold, wet, autumn and winter months. The ratio between the maximum and the minimum discharges, calculated on an average basis, is greatest for rivers in Ireland, the Scottish Highlands, Cumbria and the Pennines, Wales and Southwest England, all regions of relatively high precipitation, but least in the low-lying Midlands and southeast of England. Note the correspondence with respectively the 'Old' and 'Young' geological realms distinguished in Chs. 6 and 12. There are corresponding differences of regime as perceived on a daily basis. Rivers in areas of large discharge ratio tend to be 'flashy', rising swiftly and frequently to high throughputs, with sudden freshets and flooding as a result. Rivers where the discharge ratio is small tend to be more tranquil, varying smoothly in discharge on the short term. A critical value of the daily discharge is that described by hydrologists as 'bankfull', when the river just overtops its banks and begins to swamp the floodplains (the media's beloved 'bursting of banks'). Most British and Irish rivers are at least partly 'managed', however, and now flood only under exceptional conditions.

That river channels exhibit a continuum of forms in plan has been noted since the days of the Greek traveller Herodotus. Some are of *low sinuosity*, others are *braided*, with lozenge-shaped, changeable bars of sediment within the broad spread of water, and others again are strongly sinuous or *meandering* (Fig. 7.1). Braided rivers, and most of those of low sinuosity, typically have steeply sloping channels. These rivers also tend to have gravel beds and to be flashy. Their channels are unstable. A braided river displays *bars of sediment* that shift between numerous, repeatedly branching and rejoining subordinate channels. There is no floodplain, although an individual bar, and the fine sediment accumulated on it, may survive and gain vegetation for a few seasons before it is destroyed. In contrast, meandering streams, presenting a single channel, tend to be of low or modest discharge ratio, with comparatively low channel slopes, sandy or mixed gravel and sand beds, and wide floodplains. Their channels are relatively stable, and change only slowly as meander-loops migrate sideways through the earlier alluvium. From time to time meanders become cut off, persisting within the floodplain as *oxbow lakes* that evolve slowly into meres. Because of where they have built up, some low-sinuosity rivers have unusually low slopes, little erosive power and, consequently, stable muddy banks and beds.

During an extreme flood, a river occupying wide floodplains may abandon a long portion of its channel and cut a permanent new one to one side. This process is known as *avulsion*. It can be important archaeologically, depriving settlements on the banks of the abandoned reach of much of their economic *raison d'etre*.

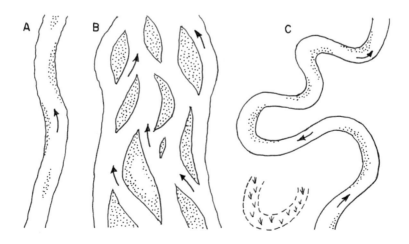

*Fig. 7.1 Schematic river channel patterns. A – low-sinuosity. B – braided. C – meandering.
Sediment accumulations not stabilized by vegetation shown stippled*

Sedimentation in rivers

The particulate sediment load of rivers consists of mineral particles and pieces of rock (see Chs. 2, 3) that are heavier than water. These are transported in two ways, as bedload and as suspended load. In *bedload transport*, restricted to gravel and largely to sand, the particles slide, roll or take very flat leaps (saltate) over the channel bed under the pressure and shear stress exerted by the flow. In the process they become rounded through impact and abrasion and sorted into like-with-like size fractions. Bedload material is therefore largely confined to the channel itself and is only exceptionally driven up onto the floodplains above. The *suspended load*, typically mud and the finest sand, is distributed throughout the body of the flow, rendering the whole river turbid. It consists of particles small enough to be more or less permanently buoyed up against gravity by the turbulent eddies of the current, which exert a net upward thrust on the relatively dense grains. The suspended load is delivered far and wide over the floodplains whenever a river floods. Hence these two modes of transport, acting within the framework of the regime, differentiate the sediment load of a river and determine where its textural elements are typically deposited.

Even where a stream has straight banks, the current winds from side to side, an effect due to viscosity that is most exaggerated in meandering river and tidal channels. The curving flow follows a *spiral path* around a meander (Fig. 7.2), the current at the bed sweeping obliquely up toward the inner bank while that at the surface crosses obliquely toward the outer side. The consequence

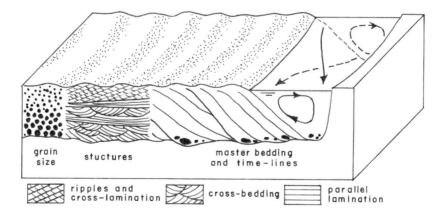

Fig. 7.2 A model for flow and sedimentation in a curved or meandering river reach

is that the bedload accretes on the inner bank, forming the meander *point bar*, while the outer bank experiences an harmonious undercutting and erosion, the channel advancing sideways without change of size or profile. The general effect is easily demonstrated. Partly fill a flat-bottomed mug with water and sprinkle in a little clean sand and some tiny shreds of paper. Gently stir the mug and remove the stirrer. The sand rolls toward the axis of the mug, while the floating shreds travel toward the sides. These flow-markers are responding to an otherwise invisible spiral current generated in the mug, due to exactly the same viscous effect that creates the spiral flow in a meander.

The sideways advance of a meander allows a thick, sandy or, less commonly, gravelly *point-bar deposit* to grow on top of an eroded surface swept out by the moving channel profile (Fig. 7.2). Such sedimentary structures as *cross-bedding*, *ripple marks*, and *parallel lamination* are likely to be present in the sands (Ch. 3), and there may be a basal, gravelly *lag* that includes lumps of cohesive soil collapsed from the outer bank. The whole deposit is described as *laterally accreted*, for major bedding surfaces, and also imaginary time-lines, are not horizontal but incline from top to bottom of the bar deposit and dip in the direction of bar migration. One archaeological consequence is that artefacts released into the river, including waterlogged wood, tend to find their way to the bottom of the channel and therefore appear at and near the base of the point-bar deposit. They may therefore be thought older than they truly are, unless evidence for lateral accretion has been recognized.

All river channels, and not just meandering ones, experience change as the result of bank erosion and the remodeling of bars. Braided rivers show these effects in the extreme. Erosion surfaces are common within and beneath fluvial channel deposits. Channel wandering tends to destroy archaeological sites but some changes, for example, meander cut-off and avulsion, may isolate them from further influence.

Formed from the suspended load, *floodplain deposits* are very different. Typically, they are muds and sandy muds with only occasional thin, sharp-based layers of sand. They display features, such as desiccation cracks, rain prints, animal tracks, plant beds, and immature soils, which point to episodic accumulation under mixed aqueous and atmospheric conditions.

Channel and water management

Most large rivers in Britain and Ireland have over the last several centuries been at least partly managed, in order to change or increase land-holdings, power mills, facilitate fishing, and improve navigation. Commonly with documentary support, the archaeological evidence for milling takes the form of, ideally, standing buildings but in many cases just traces of weirs, sluices, leats and millponds. Archaeologically, attempts at flood protection are recorded by weirs of wood or stone built across the channel, and as earthen levees, a much-neglected sort of field monument, erected along the banks. Embanked groups of fields are widely used as temporary reservoirs for winter floodwaters, as may be seen in the Fenland. Many historic weirs, however, were built for fishing and interfered with navigation but, by including a lock, allowed trading by boat to continue. The presence along a river of mounds of dredged-up bed material, suspiciously straight reaches, and frequent cut-off channel loops all point to further deliberate improvements in trafficability and to the creation of 'navigations' (canals are entirely artificial). What cannot be underestimated is the importance of water transport in the past, acknowledged since Roman times as an order of magnitude or more cheaper than carriage by road, at least up to the era of the internal combustion engine.

The most impressive evidence for *water management* comes from the Fenland Basin in eastern England, a level outcrop *c.* 4000 km² in extent of Holocene fluviomarine deposits contained between a ridge of Jurassic rocks in the west and a Cretaceous escarpment to the east. The region is famous for its great, open skies. As well as lesser ones, four major rivers – the Great Ouse, Nene, Welland and Witham – flow into this area, today of fertile agricultural land. Attempts at flood management and drainage by cutting artificial ditches or drains began in the Roman period and were pursued sporadically in late medieval times (e.g. Morton's Leam, 1478). Widespread efforts date from the seventeenth century, when Dutch engineers, in a superb example of 'technology transfer', were recruited to dig 'drains' or 'cuts', the larger named after their width (e.g. the Hundred Foot), and to reroute river channels. A monument to these and similar later efforts is the present extensive network of embanked waterways, and especially the complex of channels and carefully managed sluices at Denver Sluice (Downham Market, Norfolk).

The successful drainage of the Fens, and similar areas, however, has not been achieved without great initial and ongoing cost. The land has everywhere subsided – at Whittlesey Mere (Ch . 9) by *c.* 4 m – so that many districts now lie well below sea level, and must be continuously pumped free of water. As evidence, a windmill survives at Wicken Fen and a few steam-powered pumping stations still stand; electric have now replaced diesel pumps. There has been considerable loss of soil substance from dried out peaty areas in the inner fens, due to oxidation and wind-blow. The prevention of marine floodings has demanded the construction along the coast of many kilometres of large, expensive seabanks.

Catchments and alluviation

Precipitation falling on catchments reaches British and Irish rivers largely by slow underground flow through pores, fractures and small, natural pipes present in the soil. Especially during wet weather, another proportion arrives episodically by overland sheet flow/hillwash and, where soil erosion is occurring on cultivated fields, as temporary streamlets. The responses of catchments to changes in climate and land-use profoundly influence river sedimentation.

River discharge decreases with falling precipitation, and over time may cause a change in channel pattern from braided to meandering. Increase tends to cause deepening and widening. Vegetation in catchments, and especially a tree-cover, protects the surface of the soil, anchors soil, slows the rate of runoff, and encourages infiltration and water-retention, smoothing out variations in precipitation. Removal of this cover promotes a flashier regime and, if cultivation also follows, increases sediment yields by many times, especially if done in spring or autumn. Headwater channels may erode deeper and significant, damaging floodplain *alluviation* occur in lower reaches. Rates on British floodplains typically now measure several millimetres annually but for some rivers reach two or three centimetres. The processing of coal and metallic ores at mining sites with access to rivers, yielding particulate debris, has a similar deleterious effect but leaves in the sediments a potentially dateable record of the contamination.

The archaeological consequences are mixed. Floodplain alluviation both preserves and hides activity and occupation sites, but later channel wandering and bank erosion may reveal what became hidden. Important evidence of past industrial activities is preserved in the floodplain record of contaminants from mining sites. Coal dust from the mines and coaling ports of South Wales is now so widespread in the younger sediments of the well-mixed Severn Estuary that it can be used to date and correlate them.

Further Reading

Coles, J. & Hall, D. 1998. *Changing landscapes: the ancient Fenland*. Cambridge, Cambridgeshire County Council

Darby, H.C. 1968. *The draining of the Fens*. Cambridge, Cambridge University Press

Davies. B.E. & Lewin, J. 1974. Chronosequences in alluvial soils with special reference to historic lead pollution in Cardiganshire, Wales. *Environmental Pollution* B6, 49-57

Fulford, M.G. 1992. A post-medieval mill at Woolaston. *Transactions of the Bristol and Gloucestershire Archaeological Society* 110, 124-12

Lewin, J. 1981. *British rivers*. London, Allen & Unwin

Macklin, M.A. & Lewin, J. 1993. Holocene river alluviation in Britain. *Zeitschrift für Geomorphologie* 88, 109-122

Morisawa, M. 1968. *Streams, their dynamics and morphology.* New York, MacGraw-Hill

Needham, S. & Macklin, M.G. 1992. *Alluvial archaeology in Britain.* Oxford, Oxbow

---

Case 1: Woolaston Grange Farm – a West Gloucstershire 'hydraulic' landscape

Woolaston Grange Farm, on a Tintern Abbey holding, is a large, eighteenth-century farmhouse stoutly built of Lower Old Red Sandstone. It has stone barns and outbuildings and, to the south, a ruined chapel and earthworks. The land hereabouts slopes gently southeastward, and is drained by small streams crossing outcrops of impermeable Devonian and Triassic mudrocks, capped by Pleistocene terrace gravels, reaching the tidal Severn at a narrow saltmarsh. Water milling and management have been practiced in the area over a long period (Fig. 7.3).

Black Brook rises on the southeastern slopes of the Forest of Dean massif and flows past Woolaston Grange Farm, reaching tidewater as Grange Pill. The stream is straight and apparently canalized for *c.* 200 m near Wyvern Farm and again, in straight or smoothly curving reaches, from near Pope's Well onward. From just south of Wyvern Farm a leat, roughly following the 10 m contour, can be traced along the northeastern side of the valley to a large barn at Woolaston Grange Farm. One end houses a well-preserved, iron breast watermill used for corn during the nineteenth and earliest twentieth centuries.

About 800 m to the northeast another small stream crosses the slopes, at tidewater called Ley Pill. Antedating the railway, the lowermost 800 m is straight and clearly artificial. A double leat can be traced as two channels with slightly different elevations for *c.* 600 m along the southwestern side of the valley, to where they are lost near the head of a narrow, tapering field. As at Woolaston Grange Farm, the leats roughly follow the 10 m contour. Earthworks at the trees include two pond-like features, each associated with a channel. Roughly dressed blocks of Old Red Sandstone occur here. Excavations proved hard-standing of packed gravel and, from the upper channel, secured fragments of brick, possible late medieval pottery, a piece of a clay tobacco pipe, and wood and twigs dated by radiocarbon to the early sixteenth or early seventeenth centuries. The site is interpreted as at least one watermill.

The site of the Chesters later Roman villa and contemporary iron-making area lies in the field below the railway line to the northeast of Ley Pill. Excavation revealed many shaft furnaces, and large dumps of slag, iron ore, furnace lining, clay for furnace repairs, and bloom-refining debris. A leat heads for the complex over a distance of at least 500 m along the northeastern side of the stream valley. This feature, presumed Roman, would have secured a water supply for the villa and its industrial area.

On the left bank of Ley Pill, at Guscar Rocks, stand the remains of a substantial timber and stone quay, active from the mid twelfth to the early fifteenth century. A less elaborate counterpart of similar date occurs on the course of Hill Pill, on the opposite bank of the estuary, where Tintern Abbey also held property.

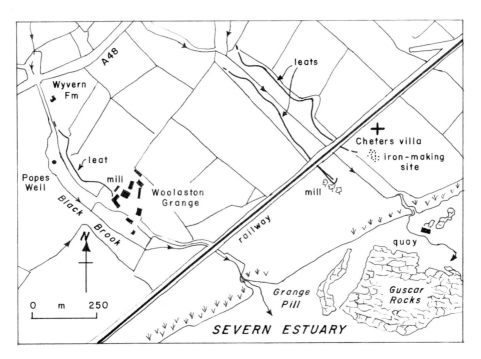

Fig 7.3 The 'hydraulic' landscape around Woolaston Grange, West Gloucestershire, on the banks of the Severn Estuary

Case 2: A medieval channel deposit in the Leicestershire Trent Valley

The mobile Trent in it middle reaches has a floodplain 2-3 km wide underlain by 3-6 m of channel gravel and sand followed by 1-2 m of floodplain silt and clay. A large gravel quarry on the river at Hemington, *c.* 15 km southwest of Nottinghsam, exposed a laterally-accreted gravel bar formed on the inner bank of a loop of the river that had migrated westward over the centuries. In the schematic reconstruction shown in Fig. 7.4, as the river flowed northward the gravel bar migrated toward the left, across an uneven, stone-strewn surface scoured into late Pleistocene deposits with clay seams and ice-wedge casts. A rich and diverse assemblage of artefacts was found on this surface and within the gravel. There were reused querns, millstones, and masonry (including Roman), two Anglo-Saxon sculptured cross-arms, two wooden post alignments, one of the ninth century AD (?fishweirs), a boat fragment, timber of the 11th and 13-14th centuries, and many grooved anchor stones. Without the quarrying these artefacts, and the activities they recorded, would have remained unknown to archaeology.

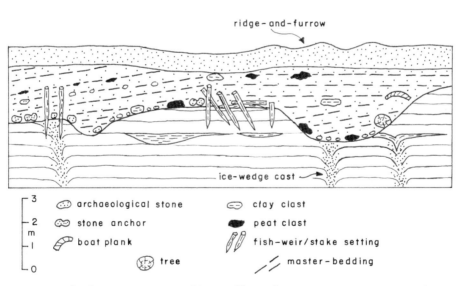

Fig. 7.4 *Partly schematic cross-section of the gravel bar on the R. Trent at Hemington, Leicstershire (adapted from C.R. Salisbury 1992)*

Case 3: Contamination from metal mining in the Tyne Valley

A granite batholith, concealed at shallow depth in the north Pennines, has mineralized the early Carboniferous sediments lying above. Veins of lead-zinc ore (galena-sphalerite) have been mined and processed at many places on tributaries in the upper catchment of the R. Tyne (Fig. 7.5). Waste material has accumulated thickly on the floodplain downstream, as shown by a 2.4 m sandy profile at Lower Prudhoe, dated mainly by historic flood events and heavy-metal tenors. Lead levels are moderately high but relatively constant at several hundred parts per million. The small amounts of silver present tend to follow the lead. Zinc levels, however, rise to several thousand parts per million, notably up to the middle of the First World War, when procurement largely ceased, and are notably 'spiky'. At several mines zinc ores were preferred over lead for a short period. One partial explanation for the fluctuating metal levels is the precarious fortunes of this extractive industry, always subject to changes in the international price of metals. Similar patterns of heavy-metal content are known from many other areas of metal mining, for example, Southwest England and mid and North Wales. The washings from coal mining similarly accumulated in river and also beach deposits downstream, as in South Wales and on Tyneside.

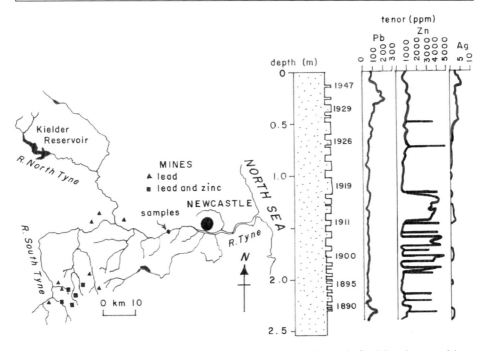

Fig. 7.5 Vertical distribution of lead-zinc-silver waste washings in silty-sandy floodplain deposits of the Tyne at Lower Prudhoe (adapted from Macklin, Rumsby & Newson 1993)

8. Sea-Level and Coasts

Coastal geological agencies

For people living in Britain and Ireland the coast is nowhere so distant that its influence has not in some way been felt. Archaeologically, the coast is a critical interface, for dwellers on it, and in communication with it, can exploit both land and marine resources and opportunities. Living on the coast is nevertheless hazardous. The coast and the coastal zone immediately to landward are subject to the vagaries of the sea. Sheltered parts of the coast are few and far between, and mainly limited to inlets and estuaries, but even here lie dangers.

Three main geological agencies shape and influence the coast: wind, waves and tides. These operate against a background of changing relative sea-level, today accelerated by global warming.

The atmosphere is a mixture of gases, chiefly nitrogen, but with a subordinate amount of oxygen, and minor amounts of water vapour, carbon dioxide and the noble gases. This mixture is of very low density compared to water, but is nonetheless viscous and capable of exerting shear stresses as well as pressure. *Wind* is the atmosphere set in forceful motion, driven essentially by the temperature gradients existing between the equator and the poles.

Where prevailing winds blow onto a coast they may pick up sand from beaches present and form belts of wind-blown dunes immediately inland. There are many such areas of coastal dunes in Britain and Ireland, some archaeologically rich.

The wind is the parent of the *waves* affecting inshore waters and lakes. When the wind blows over the sea it exerts a viscous drag on the water, ruffling the surface and creating a weak current in the wind direction. The wind is also full of turbulent eddies, and these invisible agencies push down and pull up the water surface as they sweep by, creating random surface undulations. A complicated coupling between the ruffles due to drag and the undulations due to turbulence eventually creates on the sea more or less regular, transverse waves that move with the wind. These increase in size with increasing wind speed and with the distance travelled by the wind over the sea (fetch).

The height of waves is their most important physical property so far as coastal effects are concerned, the wave-energy increasing as the square of the height. Waves posses not only kinetic energy arising from their onward motion, but also potential energy because the crests stand higher than the troughs. Wind

waves are forceful and carry large amounts of energy to the land – it is as well
to recall that one cubic metre of sea water weighs roughly one tonne – and are
the chief agents of coastal change. They cause erosional retreat in some places
but sediment accretion in others. It is much less evident that waves also act at
a distance. As waves advance, they create orbital, oscillatory currents in the
water column below that can extend down to a sufficiently near sea-bed. If
strong enough, as during storms, these currents can set bottom sediment in
motion.

Today roughly 72% of the Earth's surface is covered by water in ceaseless
motion. Responding to the same temperature gradients that create wind,
colossal streams of water different in temperature and salinity – the ocean
currents, some cold, some warm – wander majestically through the depths of
the seas. Only close to the western borders of the ocean are they relatively fast.

Another source of motion is the tripartite gravitational action of the Sun, the
orbiting Moon, and the orbiting Earth itself on the mobile ocean water. This
motion is the *astronomical tide*, the greater part of the vertical rise and fall of the
sea surface observable at a place, and the *tidal streams*, the associated reversing
horizontal flows of water thus replaced or displaced. The astronomical
tide is *multiperiodic*. On British and Irish coasts high tides and low tides are
experienced roughly every twelve-and-a-half hours. The greatest daily tidal
range – the vertical interval between high and low water – approaches 15 m
(Severn Estuary) but is nowhere less than a few metres. The range varies on
a roughly fortnightly cycle. It is greatest (spring tides) at times of new and
full moon, and least (neap tides) when the Moon is at first or third quarter.
Furthermore, the tide varies significantly on a semiannual cycle, the spring
tides rising highest around the equinoxes and least around the solstices. The
tidal streams in British and Irish inshore waters increase in speed with tidal
range. They act parallel with the adjacent coasts and are potent agents of
sediment transport. In the English, St. George's and North Channels, and in
many estuaries, the streams peak at over 1 m s^{-1}, sufficient to shift gravel, shells
and sand and mould them into bars and banks.

The *meteorological tide*, a fraction of the observed tide, arises from two
atmospheric effects. These aperiodic components are normally very small but
in stormy times are large enough to create damaging surges that flood inland.
Hydrostatic equilibrium dictates that as an atmospheric 'low' passes over British
seas local sea levels rise by roughly one centimetre for every millibar decrease
in atmospheric pressure. Persisting strong winds accompanying a low have
the effect of piling up water against nearby coasts because of drag. These two
effects acting together can lift coastal water levels in partly-confined seas by a

matter of metres above normal. Excess levels of more than 4 m were observed in places during the calamitous storm surge of 1953 in the southern North Sea.

Coastal types

A useful classification of coasts is between those that are eroding on the medium to long term and those that are accreting. The commonest type in Britain and Ireland is the eroding coast, but there are many examples of the accreting variety. Archaeological sites can be lost on eroding coasts but buried and preserved on accreting ones.

Typically, *eroding coasts* are cliffed, a response favoured by Holocene rising relative sea-levels. The average annual retreat rate varies with the geology at the coast, from a few millimetres or less for cliffs of strong, well-jointed Palaeozoic and older rocks (e.g. Scotland), to a few centimetres or a decimetre or two for those of the softer Mesozoic or Tertiary strata (e.g. Severn Estuary, English Channel), rising to a metre or more where unconsolidated Pleistocene sediments dominate (e.g. East Anglia). The process begins when wave-action undermines cliffs. In the case of strong and medium-strength rocks, undermining is followed by either the sudden, vertical collapse or toppling of columns of rock on to the beach, followed by the gradual removal of the debris by waves, and the renewal of the cycle. Unconsolidated cliffs typically fail in the form of large, arcuate, rotational slips that descend to the beach, sometimes in stages, to be slowly cleared away. The shores of eastern England in particular have suffered in this way, many medieval villages having been lost from the Lincolnshire and East Anglian coasts. *Accreting coasts*, conversely, display a net outward movement, as new sediment is added to them by the action of coastal processes, chiefly wave-induced *longshore currents*. These coasts are flat, lie close to sea level, and typically front substantial outcrops of Holocene deposits. Many are archaeologically rich. Some are barrier coasts, consisting of a chain of sandy islands and spits that shelter tidal flats and salt marshes to landward (e.g. North Norfolk).

Dune belts

Belts of wind-blown sand and dunes (Fig. 8.1) are known from many parts of the British and Irish coast, for example, St Ive's Bay, Margam Burrows (Wales), the Sefton coast (Liverpool Bay), the famous Culbin Sands or Forest (Moray Firth) in Scotland, the Northumberland coast, the North Norfolk barrier coast, and Dundrum Bay and the Bann Estuary in Ireland.. The growth of these dune belts is not yet well understood, but it is clear that many have built up episodically over periods of centuries or millennia in response to climate and perhaps sea-

Fig. 8.1 A view seaward (northwestward) across the Penhale Sands and (stabilized) Dunes at Perranporth, Cornwall. Copyright: Historic Environment Record, Cornwall Council

level fluctuations. The oldest dunes on the Northumberland coast date from *c.* 2000 BC with further sand-blow at *c.* 800 BC and *c.* AD 500-1000. The dune system of the North Norfolk barrier coast is post-Roman, dating mainly from the 16th to the 18th centuries, coinciding with severest part of the Little Ice Age. Now occluded by extensive saltmarshes, a Roman fort, *Branodunum* (Brancaster), probably supporting a naval squadron, lies on this coast.

The climate at the coast – temperature, precipitation, windiness – is variable on several time-scales. Dune-belts seem to be a response to medium-term changes with durations of centuries. Beach sand is readily mobilized when winds are comparatively strong and tend to come from similar directions. Under these conditions sand can spread for some kilometres inland from the coast and may avoid stabilization by plants. A reduction in wind strength, or a shift in the prevailing direction, can greatly reduce or even cut off the sand supply, and so allow vegetation and soil development to stabilize the once active area. A reversal of conditions may see a return to sand blowing.

The archaeological consequence of sand-blow is to bury and preserve any activity and occupation sites established in a coastal area. Settlement and cultivation may return, however, with a favourable change of regime. The sequence of deposits in a well-developed dune belt is therefore expected to consist of beach-derived sands that record sand-blow separated by dark-coloured, buried soils, with evidence for activity and settlement, that point to stability. Lengthy episodes of sand-blow and stability are not expected to be synchronous between the west and east coasts of Britain and Ireland. West coasts are most influenced by 'Atlantic' conditions, marked by prevailing westerly winds. By contrast, east coasts are most affected by a 'Continental' regime, enhancing easterly winds.

Estuaries

An *estuary* is a partly confined, sheltered body of water where a river, introducing fresh water, meets and mingles with the sea. Examples abound in Britain and Ireland, from the smallest to the large, complicated systems associated with the Shannon-Fergus, Severn, Mersey, Solway, Clyde, Forth, Humber and Thames. They present a range of sedimentary environments under the influence of wind, waves and tide: shifting sand banks and channels, marginal sand beaches and barrier spits on the more exposed shores, and marginal, intertidal mudflats and saltmarshes in more sheltered places. The vertical sequence of sediments typical of an estuary is distinctive. In low and middle positions are cross-bedded, parallel-laminated and ripple-marked sands with bands and drapes of mud. In the upper part occur interbedded silts and ripple-marked sands formed on tidal flats, followed at the highest levels by thick silts of saltmarsh origin. That the latter occur at this high level relates to the fact that only around high tide is the water still enough for mud to settle out.

Estuaries have strongly attracted settlers in every age and many have become the seats of wealthy towns and cities. They form part of the highly productive land-sea boundary. Because they are sheltered, and linked to extensive hinterlands by their rivers, estuaries are ideal locations for fishing and trading ports, and for manufacturing of all kinds. For example, tidal waters were from relatively early times impounded to power mills and other machinery. Excavations in London, as well as archival sources, testify to the expanding role of the Thames Estuary from the Roman period onwards. The growth of Bristol (Bristow) from Anglo-Norman times is later but similarly well-documented. This settlement had especially strong maritime links with Ireland, France and Spain and, later, also, the West Indies, eastern North America, and the Baltic.

Saltmarshes

Saltmarshes are a special environment to be found in most estuaries and locally on the open coast. Essentially, they are vegetated intertidal flats, formed at levels in the higher parts of the tidal range, where salt-tolerant plants (halophytes) can survive the extreme conditions of tidal inundation, re-exposure, and wetting by rain. They consist of level, vegetated platforms dissected by systems of branching, typically meandering creeks and gullies of several orders (Fig. 8.2). These channels, from tens of metres to just a few decimeters across, allow the highest tides to flood and recede from the innermost reaches of the platforms between.

In their natural state, saltmarshes in Britain and Ireland offer many attractive resources at the land-sea interface. Burrowing molluscs (e.g. cockles, clams) can be collected at low tide from the larger creeks, which can also be fished using spears, tridents and fixed wattle or basket traps. At high tide the creeks

Fig. 8.2 Saltmarshes on the north Norfolk coast. The photograph shows an area about one kilometre square, with the sea toward the top

afford a means of longshore communication by coracle or canoe.. The salt-tolerant plants create a rich pasturage, attracting animals such as wild cattle and deer that can be hunted. During summer months, when tidal levels are relatively low, semi-domesticated cattle, sheep, and pigs can be grazed on the marshes. In the same season, salt-making is possible where sea water has been trapped in natural or artificial pools. Wild fowl can be caught at certain times of year, and reeds for bedding and roofing are available.

Most of these activities are in some way *seasonal*, the resource being accessed from outside the marsh. Permanent settlement and cultivation within a saltmarsh become possible only with some means of excluding the sea.

Huge areas of tidal marsh in many British and Irish estuaries have been made habitable by the construction of seabanks (Fig. 8.3), drains, and sluices, critical earthworks that have been much neglected archaeologically. Such landscapes can be recognized by the presence of grid-like patterns of artificial drains, traces of the larger (meandering) creeks originally present, redundant seabanks, and clifflets recording abandoned shorelines. This process of embanking and draining is commonly called reclamation, but more truthfully styled *land-claim*. In the Severn Estuary alone, over 800 km² of former marsh have been

Fig. 8.3 A late medieval, now redundant seabank on the Severn Estuary coast. The coast is to the right, where the present active seabank (1960s) can be glimpsed

transformed, beginning in the Roman period and continuing up to modern times. The area, known today as the Severn Levels, now supports many farms and villages, and some industries (e.g. Avonmouth). Land-claim increases the economic potential of an area by an order of magnitude or more but, as in the Fenland (Ch. 7), is not achieved without a significant, ongoing cost. As relative sea-level in the Severn Estuary has continued to rise, it has been necessary to repeatedly widen, raise, and shorten the seabanks along the shores. A less obvious but perhaps more serious cost is that land-claim drastically curtails the flow to the sea and its fisheries of the important nutrients generated in salt marshes. Coastal erosion, forced by estuarine regime change, in many places caused seabanks to be rebuilt further inland, especially during the Little Ice Age, the process known as *set-back*.

Relative sea-level

Relative sea-level at a place is the observed difference in elevation between a defined sea-level and a fixed position on the land. Its value changed frequently and dramatically over the 2.6 million years of the Quaternary as glaciers waxed and waned. Indeed, relative sea-level has never been stable geologically, and it continues to change significantly today. Several factors are involved.

One factor, itself of several components, is known as *eustasy*, the shape and level of the ocean surface in equilibrium with the Earth's gravitational field. So far as the Quaternary is concerned, the most important influence on relative sea-level is *glacio-eustatic* change. This is the fluctuation in the mass and density-related volume of the global ocean in response to the formation and melting of especially polar ice. Fluctuations in relative sea-level of the order of 130 m mark the changes between glacial and inter-glacial periods. The *geoidal-eustatic* component is less well understood but arises because the shape of the ocean surface varies with the slow shifting of the Earth's gravitational field. Other contributions come from changes in the position of the zonal winds and major ocean currents. The total eustatic contribution to global sea-level change today is a rise of about 3mm annually that is accelerating (global warming).

The way the Earth's crust appears to 'float' in a kind of hydrostatic equilibrium on the more mobile material beneath is called *isostasy*. The formation and melting of ice-sheets disturbs this equilibrium and causes important *glacio-isostatic* changes in the level of the land. Imagine a large region of the crust to be like an inflatable pillow. An ice-sheet a few kilometres thick significantly loads and depresses the crust, which accordingly sinks beneath it, just as the pillow becomes depressed where a heavy book is placed on it. Around the margins of the ice-sheet the crust rises up into a *fore-bulge*, just as the pillow swells up around the book. Equilibrium can be regained in each case by removing the load. The crustal rocks are stiff, however, and their equilibrium is slow to be fully restored.

During the late Pleistocene thick ice-sheets covered most of Ireland, Wales and northern Britain, forcing down the crust beneath. Deglaciation allowed these regions to rise, or *rebound*, as shown in northern Ireland and Scotland by flights of raised shoreline features, the highest a few tens of metres above present sea-level. In the fore-bulge to the south and east, the land gradually sank, a process continuing today in places at a rate of a few millimetres annually (e.g. Thames Estuary). Associated with these effects due to ice loading and unloading are *hydroisostatic* changes, the depression of continental shelves as they are flooded by the deepening ocean.

Curves of Holocene sea-level change have been established for many parts of Britain and Ireland by graphing the age and elevation of such sea-level indicators as highest intertidal / just supratidal peats, formed in coastal marshes, mollusc shells from beach deposits, and assemblages of foraminifera. Unsurprisingly, no two curves from Britain and Ireland have the same form. As maybe expected, the greatest differences lie between areas of rebound and the fore-bulge.

Further Reading

Allen, J.R.L. & Fulford, M.G. 1986. The Wentlooge Level: A Romano-British salt-marsh reclamation in southeast Wales. *Britannia* 17, 91-117

Bell, M.G. 1990. *Brean Down excavations 1983-1987*. London, English Heritage

Bell, M.G. 2013. *The Bronze Age in the Severn Estuary.* York, Council for British Archaeology.

Ellis, S. & Crowther, D.R.1990. *Humber perspectives. A region through the ages.* Hull, Hull University Press

O'Sullivan, A. 2001. *Foragers, farmers and fishers in a coastal landscape. An intertidal archaeological survey of the Shannon Estuary.* Dublin, Royal Irish Academy

Harvey, C.E. & Press, J. 1988. *Studies in the business history of Bristol.* Bristol, Bristol Academic Press

Haslett, S.K. 2000. *Coastal systems.* London, Routledge

Rippon, S.J. 1997. *The Severn Estuary. Landscape, evolution and wetland reclamation.* Leicester, Leicester University Press

Roberts, M.B. & Parfitt, S.A. 1999. *Boxgrove. A Middle Pleistocene hominid site at Earlham Quarry, Boxgrove, West Sussex.* London, English Heritage

Steers, J.S. 1964. *The coastline of England and Wales*, 2nd ed. Cambridge, Cambridge University Press

Case 1: Middle Pleistocene shoreline, Boxgrove, West Sussex

A bold rampart underlain by the Upper Cretaceous Chalk Group – the North and South Downs – encircles the Weald of Kent (Fig. 4.1). On the southern face of the South Downs, overlooking a narrow coastal plain and the English Channel, lie three raised beaches cut back into the Chalk Group (Fig. 8.4A). The highest of these, at an altitude of *c.* 40 m, dating to between 524,000 and 420,000 BP, is the Goodwood-Slindon Raised Beach Cliffline. A complex vertical and horizontal sequence of much-quarried sediments occurs at Boxgrove on this feature, which was at times an activity area for early hominids, probably *Homo* cf. *heidelbergensis* (Fig. 8.4B)

Banked on a wave-cut platform and against a marine cliff in the chalk, the sequence begins with pebbly sands with chalk blocks succeeded by three cycles of beach-inshore sands, gravels and silts – the Slindon Sands – each recording a minor marine transgression followed by regression. Overlying them are the regressive Slindon Silts, a complex sequence of interbedded muds, silts and some sands with evidence for deposition on intertidal flats and incipient saltmarshes. These beds yield flint artefacts, chiefly ovate hand-axes, some utilized flakes, and much debitage that, at individual knapping sites, could be partly refitted. In association are the butchered bones of large mammals – bear, elephant, horse, bison, deer – and a wide variety of birds and molluscs, although whether any of these were also exploited is uncertain. A hominid femur was also recovered. Above the Slindon Silts come thin, organic-rich marsh deposits. The sequence is completed, and also obscured, by a thick mass of chalk pellet and soliflucted gravels interleaved with some collapsed chalk. These deposits seem to have been emplaced during the bevelling of the upper part of the original cliff.

Boxgrove records a complicated interglacial event (Oxygen Isotope Stage 13), marked by a number of sea-level fluctuations, followed by a deep glacial episode (Stage 12). The present elevation of the site probably has a tectonic cause.

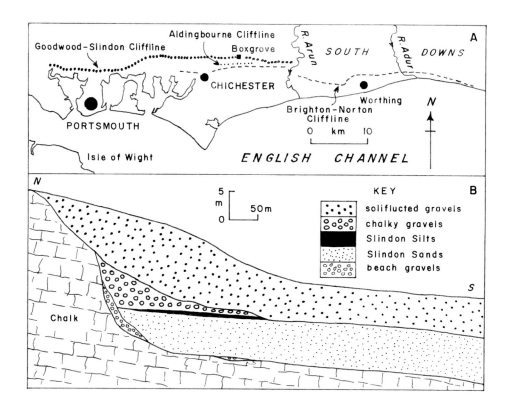

Fig. 8.4 Middle Pleistocene deposits at Boxgrove, West Sussex (adapted from Roberts and Parfitt 1999).
A – Pleistocene clifflines. B – the generalized sediment sequence at Boxgrove

Case 2: Shannon-Fergus Estuary, Central Ireland

The Shannon Estuary, with the smaller Fergus Estuary and Limerick City at its head, strikes southwestwards across the western Irish Midlands to join the Atlantic Ocean between Loop Head and Kerry Head (Fig. 8.5). The margins are rocky in places but in others formed of Holocene mudflat and tidal marsh deposits mostly the subject of last-millennium land-claim. Especially in the upper estuary, intense human activity from an early date is recorded in the Holocene intertidal sediments and on the adjoining embanked wetlands. A submerged forest of scots pine underpins the Holocene sequence and is exposed in many places in the lower and upper estuary.

The Mesolithic is recorded by a single wooden plank from the upper estuary. More extensive are signs of Neolithic activity, in the form of occupation sites, stone axes, flintwork, worked wood and baskets, hazel nuts, exploited mammal bones, and spreads of red deer bones (probably natural deaths). Bronze Age sites abound in the upper Shannon and the Fergus estuaries. They include houses, cattle and other bones, stone slabs, worked wood, wooden planks, post-and-wattle fishing hedges, and a withy tie. Many occupation/activity sites of this date are known from the dry hinterlands. The Iron Age, however, is archaeologically transparent (?climatic reasons).

Fishing on the intertidal flats and in the creeks using fixed engines was important during medieval times, both early (AD 400–800) and late (AD 1100–1350), as well as more recently. Early post-and-wattle fences, V-shaped traps, and large and small woven baskets, are all recorded in the intertidal zone. The late medieval traps were especially large and diverse: C-shaped, tick-shaped, L-shaped, stake-and-net, and transverse (in creeks), depending on where they were set. These fisheries served numerous settlements of the period in the hinterland. The medieval traps were chiefly made using oak for posts and young roundwood of hazel and some alder for woven wattle. From Norman times onwards the tidal wetlands throughout the Shannon and Fergus estuaries were embanked (corcass), episodically in places, to the substantial total of at least 115 km^2.

The outer estuary in particular remains poorly explored archaeologically.

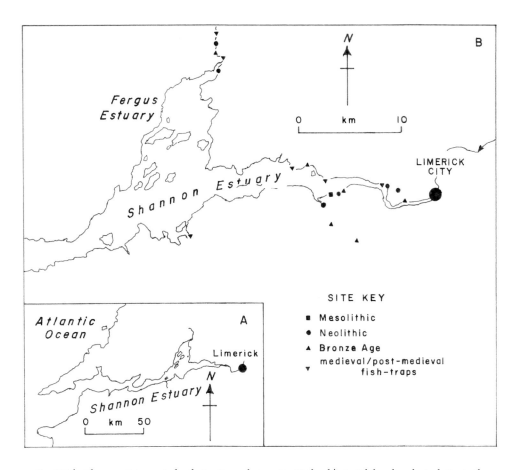

Fig. 8.5 The Shannon Estuary, Ireland. A - General view. B - Wetland/intertidal archaeological sites in the upper Shannon and Fergus Estuaries (generalized from O'Sullivan 2001)

Case 3: Brean Down (Burnham Dunes), Somerset

Brean Down is the southernmost of three bold peninsulas of Lower Carboniferous Limestone that project westward into the Bristol Channel from the Somerset coast. It lies at the northern end of a belt of blown sand that ranges southward for 10 km to Burnham-on-Sea. The sandcliff and the slopes below immediately south of the Down display a short sequence of Pleistocene deposits followed, on a buried soil, by a thick accumulation of Beaker, Bronze Age and Iron Age blown sands. The Beaker sands yielded representative pottery, and charcoal, bone and flintwork that could include some that was Neolithic. The Bronze Age sand and clay contained stone structures, including house sites, and afforded much pottery, charcoal, flintwork, marine molluscs, and salt-making briquetage, pointing to the lengthy presence of a vibrant community. There was also evidence amongst the sands for Iron Age and medieval activity and for a sub-Roman cemetery. On the Down itself Celtic fields, a Romano-Celtic temple, and a modern fort have been found.

9. Stone for Building

National resources

Britain and Ireland are perhaps unique in the sheer range and diversity of their geological makeup. Rocks and strata of almost every date from over three billion years ago to the present are crammed into these islands lying between mainland Europe and the Americas. A huge variety of building stone has consequently been available. Because of the rapidity with which outcrops change over the ground (e.g. Fig. 4.1), an extraordinary regionality is recognizable in the appearance of even modern stone buildings and monuments. Because rock – heavy and bulky – was rarely carried far, the national geological maps can be largely reconstructed by simply mapping building stone. Slate-rich Welsh or Cumbrian settlements are unmistakably different in aspect from lava-rich Northern Irish towns, granite-dominated northeast Scottish cities, sandstone-faced Pennine cotton and wool towns, and flint-built East Anglian villages.

The right stone?

It all depends on the purpose intended for the stone. As the Roman architect Vitruvius would have agreed, in making a choice, ready availability, workability and strength, resistance to weathering, and pleasing appearance could all be important.

Ready availability implies not only the nearby presence of exposures or quarries of the desired stone but also ease of access for means of transport, whether carts or boats. *Workability* depends on the geological character of the stone, its composition and crystal/grain size, hardness, jointing, bedding, and lamination or cleavage. Rocks such as granite and gabbro are coarsely crystalline, hard, widely jointed and without bedding. They are difficult to shape but can be formed into large, monumental blocks, but not moulded other than crudely. In contrast are sedimentary rocks such as the Lower Old Red Sandstone of Herefordshire and the Scottish Borders, and the Middle Jurassic oolitic-shelly limestones of the Cotswolds and East Midlands, especially generic Bathstone. These rocks, of a fine, even grain, are *freestones* that can be worked in any direction, and moulded using chisels and files into delicate, deeply undercut shapes. They are isotropic in terms of their mechanical properties. Most sedimentary rocks, however, are anisotropic, like wood. They may be thickly bedded but internally possess laminae that make the rock much stronger across the plane of the bedding than parallel with it. Such rocks can be readily split into slabs, flags, and tiles. If the joints are widely spaced, as in the Devonian

Caithness Flagstone Series of northeast Scotland and Orkney, thin plates two metres square or more can be quarried. The ultimate anisotropic rock is slate, found widely in 'Old' Britain, the finest of which can be split into sheets just a few millimetres thick for roofing, as in Cornwall and Wales.

An appreciation of the *weatherabililty* of a building stone is generally a matter of experience over time. The latter is not always available and in the past many poor choices, now the bane of modern conservationists, have been made. Weatherabililty depends mainly on composition and the presence of internal partings and voids that frost or lichens can exploit. Moreover, it has never helped to lay stone with the bedding vertical. Calcite-cemented limestones and sandstones, particularly if also porous and slightly clayey, are especially prone to weather rapidly. Over time blocks become rounded at the edges and corners, and large flakes may spall off shaped faces. By contrast, rocks such as granite and gabbro have a marked resistance to weathering.

Whether a building stone is of a *pleasing appearance* when either fresh or weathered is generally of secondary importance, but native stone of many colours is available in Britain and Ireland: white (early Cretaceous Kentish rag), grey (Carboniferous Limestone, Pennant sandstone); cream-yellow (Carboniferous York stone, Triassic Sudbrook Sandstone, Jurassic Bathstone and Headington stone); grey-green (Ordovician Borrowdale Volcanic Series, Cretaceous Upper Greensand), pink (Devonian Lower Old Red Sandstone); red (Triassic Hollington stone, Mansfield stone); brown (Lower Jurassic Ham Hill stone and Banbury Ironstone, Cretaceous Norfolk carrstone); and black (some Lower Carboniferous Limestone, some Welsh and Scottish Border slates).

Not all building stone is necessarily quarried or levered off outcrops. What might be called 'found stone' is simply collected as rounded pebbles and cobbles from stream beds and beaches or garnered from fields. Much flint used in southeast England was either dug from gravels (rounded, water worn), carted from beaches (rounded), or picked from fields (angular, frost-shattered).

Styles and applications

It is important to distinguish between the character of the individual blocks of stone in a masonry building and the appearance of the collective. Individual blocks may be either *naturally shaped*, *random*, or *dressed*. The shape of a natural block depends on the geometry of the system of joints and bedding in the rock. A random block has an irregular shape as a result of breaking up stone at the quarry (Fig. 9.1A). On the other hand, dressed blocks have been deliberately shaped into right-angled forms using hammers and chisels or saws. The extent and quality of the dressing varies from rough to fine. Blocks with barely treated or flaked surfaces are described as *rock-faced* (Fig 9.1B, C). At the fine end are blocks of *ashlar*, with smooth, flat faces and shapes so accurately fashioned that, even when very large, the blocks fit together with millimetre precision and seem at first-glance unmortared (Fig. 9.1D, E).

A low masonry wall may be built using rows of random or roughly-dressed stone a single block deep with mortar between. More substantial walls call for a different approach. A common style is a facing of worked stone on the outside and inside separated by a rubble core, provision being made at intervals for lacing stones (Fig. 9.2). The core in some cases is a random mixture of broken stone and mortar, but in others consists of carefully laid stone. Dressed stone outer facings are made either of coursed, largish to large ashlar or of relatively small dressed blocks, in the style known as *petit appareil* (Fig. 9.1D). Inner facings are usually of either roughly dressed or random blocks. Walls of simply coursed stone are straightforward to make but can seem dull. Nineteenth-century builders overcame this limitation with *snecked* walls of carefully fitted blocks of different sizes (Fig. 9.1F). Tall, thick walls, such as fort and town walls in Roman Britain, were usually erected in successive lifts each about a metre thick, capped by a layer of either bricks or stone slabs that stretched as a lacing across the full width (Fig. 9.3). In a style called *shingling*, introduced in Roman times and used into the early medieval period, the facing/core is of coursed, tilted blocks mortared together (Figs. 9.3, 9.4). A course of shingled stone clearly has an orientation, recording the direction in which the work-gang built, either to the left or the right of an observer.

Bricks became an increasingly important building material from the late medieval period onward, especially in stone-poor areas (Ch. 11). They can be laid in many ways to make either solid walls or rubble-filled walls, to either of which stone facings may be attached, a common early modern and modern practice. The facings encountered range, for example, from Bathstone ashlar to rounded flint pebbles or cobbles, carefully selected to be of a similar size.

A distinctive kind of apparently unmortared facing is known from regions where rock is available in thin plates, such as Cumbria and Wales (coarse slates), and western Norfolk ('small' carrstone). The plates are laid horizontally, with a slight outward dip to reject rain, the mortar being placed in a concealed strip to the rear (Fig. 9.5).

Stone has many other uses than for walls. Thinly laminated sandstones and siltstones have been quarried for centuries to create large, rectangular flagstones for floors and pavements. Hard standings and rough floors can be made by setting upright or *pitching* platy fragments of stone, either quarried

Fig. 9.1 Some buildings styles. A – random stone, chalk. B – rock-faced stone, oncolitic limestone. C – rock-faced stone, Pennant sandstone D – ashlar, Bathstone. E – ashlar, Bathstone. F – snecked, carrstone.

or found, and also bricks. (Fig. 9.2). In British and Irish urban environments, especially from the early nineteenth century, stone was used for making kerbs and gutters and for road-making generally. Widely cast imports were necessary where a town lay in a stone-poor region. At Reading, Berkshire, for example, such stone was imported from Wiltshire, Bristol, the Welsh Borders, the East Midlands, and even as far away as the Channel Islands and Norway.

Decorative stone

Stone declared to be decorative is aesthetically pleasing in colour, texture and pattern. Such material is used architecturally for walling, columns, and statuary,

Fig. 9.2 Rubble-filled walls and pitched stone floor of an early modern building at Dolaucothi-Pumpsaint southwest Wales. Photo: B. Burnham

Fig. 9.3 External view of part of the town wall at Roman Silchester showing the exposed, shingled wall core of flint and layers of lacing stone slabs

Fig. 9.4 Shingled flint facing, Sts. Peter and Paul, Checkendon, South Oxfordshire

Fig. 9.5 House wall of small carr, Norfolk

and internally for wall panels, geometrical floor tiles and mosaic tesserae, inlay work, and fonts, pulpits, and church altar-pieces and monuments.

Perhaps the most important of decorative stones are the *marbles*, a term extended from metamorphosed limestones to any calcareous rock taking a polish. The former are restricted in Britain and Ireland to the lightly metamorphosed, mainly grey, pink or red Devonian limestones of south Devon and the black Carboniferous limestones of southern Ireland. The latter include the grey Rhaetic marble of Somerset, with its decorative 'forests', and the long-quarried, grey, gastropod-rich Purbeck marble from the Upper Jurassic of coastal Dorset. Visit Westminster Abbey for a forceful demonstration of the varied use of the last.

Mortars and plasters

Mortar spreads the load of the stone blocks or bricks used to make structures, and also binds these elements together. Classically, British and Irish mortars are *lime mortars*, made by mixing slaked lime ($Ca(OH)_2$) with water and sharp sand to form a thick paste, to which a little wood or coal ashes, fine gravel, or crushed brick (e.g. Roman *opus signinum*) may be added as desired.

Making slaked lime is a simple process. First, a calcareous rock such as chalk is calcined in a clamp or kiln using a suitable carbon fuel. The calcium carbonate is broken down to form carbon dioxide, released to the air, and quick lime (CaO), for sale to building sites. Here slaked lime is made by adding water to quick lime held in a pit or large tub. In recent times, specialized mortars were made by calcining cementstones, such as occur in the London Clay Formation of the Thames Estuary. These are concretions combining calcium carbonate with clay minerals that are partly converted to cement-like calcium silicates when burnt. Of modern introduction are *cement mortars*, another bane of conservationists.

Lime mortars harden in two stages. The first, rapid stage is the drying off of water. In the second, much slower stage the slaked lime gradually combines with atmospheric carbon dioxide to recreate calcium carbonate, completing a chemical cycle. The end product is a strong, porous, rock-like material that allows the masonry to 'breathe', something not permitted by cement mortars.

Typically, plasters are used to smooth an internal wall. *Wall plasters* are mixtures in water of plaster of Paris ($CaSO_4$), sharp sand, and a fibrous binder such as horse hair or wool. Plaster of Paris is obtained by burning gypsum, hydrated calcium sulphate ($CaSO_4.2H_2O$), found in British Triassic rocks. The plaster is applied rapidly in layers, beginning with a sandy, fibrous mix, and ending with a thin skin of almost pure plaster of Paris.

Further Reading

Adam, J.-P. 1994. *Roman building. Materials and techniques.* London, Routledge

Allen, J.R.L. 2009. The geology of early Roman mosaics and *opus sectile* in southernmost Britain: a summary. *Mosaic* 36, 5-10

Allen, J.R.L. 2012. The defences of *Venta Silurum* (Caerwent): a new analysis of the building programme. *Monmouthshire Antiquary* 28, 3-32

Allen, J.R.L. 2015. *Kerbside geology in Reading. Aspects of the historical archaeology of the expanding town c. 1840-1914.* Oxford, Archaeopress

Allen, J.R.L.2016 (April). Stone industries over time. *Geoscientist,* 26(3), 3

Brunskill, R.W. 1990. *Brick building in Britain.* London, Gollancz

Childe, V.G. 1931. *Skara Brae. A Pictish village in Orkney.* London, Kegan Paul

Godwin, C.G. 1984. Mining in the Elland Flags: a forgotten Yorkshire industry. *British Geological Survey Report 16*

Jope, E.M. 1964. The Saxon building-stone industrry in southern and Midland England. *Medieval Archaeology* 8, 91-118

Morgan, N. & Powell, P. 2015. *The geology of Oxford gravestones.* Oxford, Geologica Press

Pearson, A. 2006. *The work of giants. Stone and quarrying in Roman Britain.* Stroud, Tempus

Potter, J.F. 2009. Decorative Anglo-Saxon wall fabrics: a fashion in stonework. *Church Archaeology* 13, 41-51

Price, M.T. 2007. *Decorative stone.* London, Thames & Hudson

Stanier, P. 2000. *Stone quarry landscapes.* Stroud, Tempus

Williams, R. 2004. *Limekilns and limeburning.* Shire, Princes Risborough

Fig. 9.6 The (restored) interior of hut 1 at Skara Brae, Orkney

Case 1: Skara Brae, Orkney

In mid Devonian times eastern Scotland and the Northern Isles lay near the western margin of a vast lake that varied greatly in size as the warm climate fluctuated. The calcareous, silty muds and occasional sands that accumulated in this lake now form the Caithness Flagstone Series, a thick sequence of thinly bedded and laminated, well-jointed mudrocks that can be easily split into neat plates and slabs of a wide range of shapes and sizes and with at least one straight edge.

Neolithic people were very active in Scotland and the Northern Isles. At the Bay of Skail, in the western lowlands of Mainland, Orkney, they built in several stages the settlement now called Skara Brae, an integrated complex in stone of at least eight round or rectangular huts of different sizes linked by covered passages (Fig. 9.6). The materials are flags and slabs retrieved from the gently dipping Caithness Flagstone Series exposed in natural quarries on the nearby coast. The builders exploited to the full the unique bedding and jointing of these rocks. The unmortared walls rest on footings of slabs laid on edge and therefore well drained. Some walls, mainly internal, are single rows of narrow slabs laid horizontally on the bed. Double rows form external walls. Internally, long slabs make the jambs and lintels of the low doorways. Living quarters are furnished with box-like beds of large, upright, rectangular flags, and also cupboard- and shelf-like storage structures, made from long slabs. Some roofs seem to have been corbelled, a technique to which the stone particularly lends itself, but whalebone rafters may have otherwise been used.

Case 2: Roman Caerwent, southeast Wales

The small Roman town of *Venta Silurum* (Caearwent), in southeast Wales, on the main Roman road to the west, is distinguished by the survival of a stout masonry wall some 3 m thick at the base that dates from *c.* AD 330. Of rectangular plan with bastions, it was built almost entirely of stone from outcrops of the Carboniferous Limestone Series close nearby. The rocks exploited are moderately thickly bedded and closely jointed limestones that can be split into thin slabs as well as quarried as substantial blocks.

The wall facings (Fig. 9.7) are regular courses of mortared limestone blocks (*petit appareil*). There are rows of putlog holes at a vertical interval of *c.* 1-1.5 m, but otherwise no external evidence that the wall was built in stages with lacing stones (*cf.* Silchester, Fig. 9.3). By contrast, the wall core, widely exposed by robbing, is formed of shingled limestone slabs laid in mortared courses exactly aligned with those of the facing blocks. These courses are alternatively right-shingled and left-shingled in complex arrangements of single rows and groups that vary greatly in length and pattern from one section of the wall to another. Lateral changes in shingling direction are denoted by marker blocks, singly or in stacks. Within the constraints of the stone brought to the site, the work-gangs on the wall evidently had considerable freedom to vary their mode of building.

A

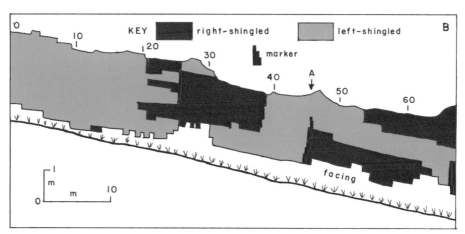

Fig. 9.7 The masonry wall of Roman Caerwen, southeast Wales. A- External view of portion of west wall c. 45 m south of West Gate. B – Reconstruction of shingling directions in the core of the west wall, including the section illustrated in A

Case 3: Barnack stone

A few centuries elapsed after the Roman period before significant building in stone was resumed in England during Anglo-Saxon and Anglo-Norman times, and then chiefly for churches. In addition to mainly local material for facings, strong freestones capable of being moulded were required, chiefly for quoins, pilasters, and doorway and window dressings. One stone extensively exploited was a coarse, cream-coloured, ooliic-shelly limestone from the Middle Jurassic Inferior Oolite beds at the village of Barnack, some 12 km northwest of Peterborough (Fig. 2.2E). The shallow, ancient quarries are marked today by a landscape known as Hills and Holes.

Barnack stone was dispersed widely throughout the Fenland Basin, the East Midlands, and East Anglia, and even reached London and Kent (Fig. 9.8), apparently chiefly by water on the numerous rivers and drains of these low-lying areas. Several kilometres to the east of Peterborough lies Whittlesey. Up to the middle of the nineteenth century, when drainage of the eponymous Mere began, the village was a small, thriving port on the old course of the R. Nene, which passed through this huge, shallow lake. Drainage revealed many artefacts on the lake bed, including blocks of Barnack stone (preserved at nearby Engine Farm), presumed to have been the cargo of a wherry or raft that foundered in a storm on its way to a building site.

Barnack stone's main competitors were other Middle Jurassic, oolitic-shelly limestones, such as Taynton Stone, which came in from the Cotswold area to the southwest.

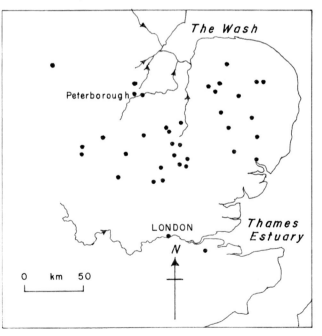

Fig. 9.8 The distribution of C8-11 buildings with Barnack stone in east-central England (adapted from Jope 1964)

Case 4: Oxford gravestones

An insight into the development of the monumental stone trade in the south Midland city of Oxford over the last three centuries or so is provided by a recent survey of gravestones in six burial grounds (Fig. 9.9).

As early as the fifteenth century, grave monuments of local shelly limestone were çarved from the Forest Marble (Middle Juraassic) at Witney and the Corallian (Upper Jurassic) beds at Headington in the northeast of the city.

Regionally-sourced Banbury ironstone (Lower Jurassic), a ferruginous shelly limestone, appeared in the earliest nineteenth century. At about the same time Pennant sandstone (Upper Carboniferous) from the Bristol area or Forest of Dean put in an appearance (Fig. 2.2D). This grey lithic sandstone appears plentifully at roughly the same time as kerbing of the same stone at Reading, Henley-on-Thames and Oxford itself, as well as in many north and east Berkshire churches and some country houses.

Several materials of regional to international provenance achieved popularity in Oxford burial grounds from the third and fourth quarters of the nineteenth century. Lower Carboniferous crinoidal limestone has a fleeting presence, but is more important as a decorative stone for internal use, as witness the early Victorian All Saints Church in Camden (London). Upper Jurassic Portland limestone was also used, probably imported by sea and then rail. Italian white marbles from Carrara appear in a small way as does Welsh slate. Granites from several sources – Peterhead, Aberdeen, Shap, Southwest England, Finland, Norway – were popular, owing to their resistance to weathering and decorative qualities when polished. Recent decades have seen the use of gabbro (?Scandinavia), Cumbrian green slate (Ordovician, Borrowdale Volcanic Series), Upper Cretaceous Nabresina limestone (Adriatic), and various gneisses and migmatites (?Scandinavia).

This sequence clearly points to an accelerating shift in procurement over time from local, through regional and national, to international resources, as means of communication developed and transport costs fell. Another, partly-matching, sequence of introduced stone, this time for kerbing and guttering, can be seen in Reading's nineteenth-century streets.

DATE / STONE	Forest Marble	Banbury Ironstone	Pennant Sandstone	crinoidal limestone	Portland Limestone	Carrara Marble	Welsh slate	granites	gabbro	Cumbrian Green Slate	Nebresina Limestone	gneisses & migmatites
>2001								2		1		1
1976 – 2000		1					1	1	3	2	3	3
1951 – 1975					1	2	1	3	1			
1926 – 1950			1		3	2	1	7				
1901 – 1925		2		1	2	2		4	1			
1876 – 1900			1	2	2	2		6				
1851 – 1875		1	2		1	1	1	1				
1826 – 1850	1	1	2									
1801 – 1820			3	1								
<1800	5											
STATUS OF SOURCE	local	regional	regional	regional	regional	international	national	international & national	international	national	international	international

Fig. 9.9 Rock-types used in Oxford burial ground and their dates of appearance (the figures are the numbers of monuments of each type)

10. Stone for Tools and Implements

Cutting and allied applications

Prehistoric peoples living in Britain and Ireland turned to flint, chert, obsidian, pitchstone and greenstone when tools with sharp and durable edges for cutting, felling, scraping and piercing were needed, and when projectile weaponry was required. Flint and chert are forms of secondary silica; obsidian and pitchstone are extrusive and intrusive igneous rocks of acid leaning toward intermediate composition. Siliceous greenstone is highly localized and very rare in Britain. These materials are all fine-grained and, like glass, are characterized by an often spectacular conchoidal fracture.

The distinctive forms of microcrystalline silica called *flint* occur in the higher parts of the Upper Cretaceous Chalk Group, cropping out in Northern Ireland, where igneous activity has baked the rocks, and in eastern and southeastern England. The smallest natural forms of flint replace the infilled dwelling and feeding burrows of invertebrates that lived in the chalk ooze. Other smaller forms replace or shroud sponges and sea-urchins. The largest bodies of flint – those suitable for tool-making – range from substantial, smooth but uneven and irregular nodules to continuous, thick bands parallel with the bedding. As attested by three-dimensional fossil preservation, flint formed at a very early stage in the burial of the chalk ooze, apparently as the result of the sulphate reduction of organic matter. Local variations in the porosity, permeability, and microchemistry of the water-logged mud strongly guided the process of silicification.

Flint is variable in hand-specimen and microscopically. Most is dull and opaque, pale to mid grey in colour, and with paler diffuse patches or well-defined mottles. Under the microscope such flint displays an intermediate to coarse microcrystallinity. The finest-grained flint is glassy, in bulk either dark grey to black or honey-coloured, and in thin flakes translucent or even transparent. This is the most prized material for tool-making, yielding strong flakes with razor-sharp edges. Flint was mined extensively during the Neolithic in southern and eastern England, but to a lesser extent in northeast Ireland. A fine example is the complex of more than 400 bell-pit mines with radiating galleries at Grimes Graves in Norfolk (Fig. 10.1). The miners here thought three-dimensionally like modern geologists, tracking the desired flint bands (bottomstone), fine-grained and very dark grey to black in colour, from surface outcrops into the shallow subsurface.

Fig. 10.1 A reconstruction by an English Heritage artist of the operation of a flint mine at Grimes Graves in Norfolk. The inset shows the plan of the interlinked galleries at a group of mine shafts (S). Copyright: English Heritage.

Chert is the name given to nodules of flint-like secondary silica found at geological levels other than the Chalk Group. The most important are in the Carboniferous Limestone Series of the Peak District, Mendips and South Wales. Exploited to some extent for tool-making, these nodules commonly display the textures – shelly or oolitic – of the limestones that were replaced. Usable chert also occurs in the Upper Jurassic of the Dorser coast. A less flint-like and visibly sponge-rich form of chert, also used for tool-making, marks some Lower Cretaceous sandstones in Wessex and the Weald.

In sharp contrast, *obsidian* and *pitchstone* are natural glasses of igneous origin found as either lavas or intrusive dykes, notably in the Tertiary volcanic districts of the Scottish Western Isle and, in a small way, on Lundy Island in the Bristol Channel. These rocks are vitreous, black and brittle, fracturing to yield razor-sharp edges. The two are different in that pitchstone includes small phenocrysts or feather-like crystal growths and may show evidence of flow. Both were used regionally for tools.

Tools were fashioned from these materials in two main ways: percussion-flaking using hard hammers or hammers and soft punches, or pressure-flaking. Axe-blanks could be smoothed by rubbing them with or against an abrasive surface, such as a piece of sandstone or, as in the chalk downlands of Wessex, a sarsen. The same procedure allowed a stone axe to be resharpened.

Sharpening

With the appearance of metals in the Copper and Bronze Ages came the incessant need for implements for sharpening articles made from them, although many stone axes of an earlier period had been similarly smoothed and finished. These unassuming but necessary implements are *whetstones* and *hones*, common finds little regarded archaeologically, but capable of affording clues to stone-based industries and trade. Whetstones are fine-grained rocks used especially with water to sharpen edge-tools such as knives, scythes, and swords. The term hone or oilstone implies a finer-grained rock still, used with oil to give razors and scalpels an especially keen edge. As there is generally no evidence of the lubricant used archaeologically, it seems appropriate to apply the term whetstone to all.

Whetstones from archaeological sites are of two main kinds according to use. One sort is a narrow *bar* of stone of rectangular section up to 30-35 cm long that is held in the hand and drawn backward and forward along the blade being sharpened, such as a sword or scythe. Such tools were made by grooving slabs of sandstone and then breaking off bars like pieces of chocolate. The other type is a large, thin, *tablet* of stone, held stationary on a knee or bench, over which the item – for example, a kitchen knife or mason's chisel – is repeatedly drawn. Whetstones may also be seen as merely the portable forms of natural rock exposures that had otherwise been exploited for sharpening and polishing, as in the case of Neolithic axes finished on Wiltshire sarsens. Once building in stone had begun, standing masonry and discarded roofing tiles were also used for sharpening, a practice that housewives at their doorsteps carried forward into modern times.

What makes rocks good whetstones? The essential requirement is a mixture of soft and hard components. The hard components, such as quartz sand, act as tiny chisels during the relative motion of whetstone and implement, removing minute slivers of metal with each pass. However, hard grains can act in this way only if the sharp edges of the particles are emergent from a background that is softer, such as calcite cement or a clay-mineral matrix. In other cases, the 'soft' background takes the form of voids, such as the vesicles in lavas or the residual pores of partly cemented sandstones. The texture of the rock strongly affects the quality of the edge produced by sharpening, because it controls the size of the metal slivers produced. Unsurprisingly, the keenest edges are yielded by the finest-grained rocks such as slates.

Although few systematic studies have so far been attempted, chiefly of medieval items, it is clear that whetstones had many sources. In the medieval period Scandinavia exported large numbers of manufactured whetstones of fine-grained metamorphic rocks. Previously whetstones were chiefly sourced within Britain and Ireland. Various sandstones, siltstones and limestones, and even igneous rocks such as basalt and dolerite, were pressed into service. Many whetstones seem to have been opportunistic finds of a suitable size, shape and lithology retrieved from cultivated fields or picked up from beaches and rivers. In stone-rich parts of the country one did not need to search far. During Roman times stone roofing tiles imported into towns were used as whetstones after perhaps some reshaping, mainly after buildings had been demolished. Although smoothed by use, many such whetstones still retain the original fixing holes for nails, a clear proof of their secondary origin. Unlike the rotary quernstones discussed below, deliberately manufactured whetstones are relatively uncommon. One Roman industry, providing whetstone bars of Old Red Sandstone, but only for the local legionary market, is known from first-century AD Usk in South Wales. Slabs of sandstone were grooved on both sides, allowing bars to be snapped off. Of much greater importance is a long-lived industry that has recently come to light based on earliest Cretaceous sandstones found in the northwest Weald. Its bar-shaped products, made in much the same way as at Usk, reached far into Roman Britain and even the near-Continent.

Grinding and milling

Bread, the staple of European peoples, can be made only after cereal grains have been ground to flour in some kind of milling device. Four kinds were used in Britain and Ireland before modern times: hand-operated *saddle querns*, hand-operated *rotary querns*, animal- or slave-powered *Pompeiian* hourglass or 'donkey' *mills*, and animal-powered or mechanical *rotary mills.* An increasing scale and efficiency of production over time is denoted by this sequence of types. The items have all been the subject of much but not yet sufficient systematic study, through petrographic, distributional and, in a few cases, geochemical analysis. Saddle querns were in use from Neolithic times until at least the mid Iron Age. Rotary querns range abundantly from the Iron Age through the Roman period. Pompeiian mills, all of lava, are known from Roman Britain, but are very rare. Combined petrogaphic and geochemical analysis has traced them to Mediterranean areas. Dominant in medieval and early modern Ireland and Britain were rotary mills powered mainly by wind or water (river/tide). Today, in late modern times, flour is made commercially using roller-mills.

Saddle querns and rotary querns each consist of a lower and an upper stone. In the former the lower stone is typically an elongated slab over which the smaller, upper stone is drawn backward and forward across cereal grain introduced between them. The two stones of a rotary quern are circular and of roughly the same size and shape, normally less than 0.45 m in diameter. A handle is fitted to the upper stone so that it can be rotated over the lower, either through a restricted arc or repeatedly through a full circle. The upper stone, flat or beehive-shaped, has a slightly concave undersurface and a central hole or hopper through which grain can be introduced as needed. In some models a spindle was passed through the hopper so as to engage with a depression in the lower stone. Donkey mills, typically hourglass-shaped, comprise a rotatable upper stone, in the form of an inverted cone pierced at the top, that fits snugly over a slightly smaller, but similarly shaped, stationary lower stone. There was a device to vary the spacing between the stones and so the fineness of the flour; the upper stone carried levers to the source of power. The Biblical strongman Samson, having lost his hair and strength to Dalilah, was put with slaves to power such a mill. Rotary mills are essentially large rotary querns. They employed two, heavy, carefully patterned, circular stones a metre or so in diameter, with means of controlling the space between and the speed of rotation.

What happens during *milling*? Cereal grains are hard but much softer than the milling stones. A cereal grain newly introduced into a moving mill rapidly experiences, first, a degree of crushing under the weight of the upper stone, then rotation and flattening with the development of spiral, internal fractures and, finally, comminution as the broken parts are separated and smeared out across the asperities on the surfaces of the stones.

What makes a good milling stone? The first criterion is hardness, which makes for a low rate of wear, which in turn ensures minimal tooth-abrading grit in the flour produced. Also desirable are the same hard-soft contrasts between components found in whetstones, especially those of vesicular lavas or sandstones with an incomplete cement of secondary silica. Calcite-cemented sandstones, however, are too soft overall to make good millstones. Siliceous quartz and flint conglomerates have been successfully used for millstones, because minute facets develop on the surfaces of the clasts as they wear.

How milling stones were made depended in the character of the rock chosen and the complexity of the desired form. Saddle querns necessitated little more than shaping a sandstone slab to a parallel-sided form either by flaking using hammer and chisel or, in the case of the toughest rocks, pecking with a mason's point or light pick. Larger querns seem to have been shaped by a combination of flaking, chiselling, and pecking.

Several millstone industries, and some factory sites, have now been identified in pre-modern Britain and Ireland through careful field and petrographic work. The native industries exploited at various times the Quartz Conglomerate (Upper Old Red Sandstone) of the Welsh Borders, the Upper Carboniferous Millstone Grit in the Pennines, the Hythe Beds (Lower Cretaceous) of the southern Weald, and the Hertfordshire Puddingstone (Pleistocene) of the southeast Midlands. The imports that flooded in from time to time included German lavas in the later Roman Period and highly silicified limestones (burrstones) from the Continent in medieval times. Rotary querns of Alderney Sandstone were imported from the Channel Islands to some Romano-British sites.

Further Reading

Allen, J.R.L. 2014. *Whetstones from Roman Silchester* (Calleva Atrebatum), *North Hampshire*. Oxford, Archaeopress.
Bradley, R. & Edmonds, M. 1993. *Interpreting the axe trade. Production and exchange on Neolithic Britain*. Cambridge, Cambridge University Press
Mason, H.J. 1978. *Flint, the Versatile Stone*. Ely, Providence Press.
Shaffrey, R. 2006. *Grinding and Milling*. Oxford, Archaeopress.
Williams, D & Peacock, D. 2011. *Bread for the people: the archaeology of mills and milling*. Oxford, Archaeopress

Case 1: Neolithic axe production at Great Langdale, Cumbria

Stone axes of many geological origins were produced and circulated in huge numbers in Neolithic Britain and Ireland, both for practical use and as prestige items. Perhaps the largest production centre was high in the mountains of Great Langdale, in the heart of the Cumbrian massif. Roughouts and polished axes from this source have a dense distribution in Britain, from the Moray Firth in Scotland to southern England. The stone, exploited at several sites, is a fine-grained, epidotized tuff stratified in the Ordovician Borrowdale Volcanic Series with similar mechanical properties to flint, breaking with a conchoidal fracture. Most production, from shallow adits, quarries, and scree, took place at almost inaccessible places amongst the misty crags high in the mountains of Great Langdale (Fig. 10.2). Equally suitable stone occurred at readily accessible sites at much lower altitudes, as an incisive petrophysical analysis showed, but it was never worked. Clearly, stone from the heights was regarded as having special virtues and perhaps magical powers.

Fig. 10.2 A Neolithic quarry for axe-blanks with a scree of trimming debris high up in the Langdale Pikes. Photo: Mark Edmonds

Case 2: Roman Wealden whetstone industry

During excavations in the 1920s at the Roman town of *Uriconium* (Wroxeter), in the Welsh Borders, an extraordinary discovery was made at the site of the forum. Concealed in a ditch beneath building rubble were about 100 unused, bar-shaped whetstones, made from a greenish grey, slightly fossiliferous, fine-grained, calcareous sandstone, probably tumbled after a fire from a shop or stall or a merchant's store. Accompanying them were substantial nests of mortaria and plain and decorated samian ware. The blank whetstones bore conspicuous, rectangular rebates along each long edge, evidence of their manufacture by snapping the ends off sandstone slabs that had been grooved across opposite faces.

The whetstones were examined by specialists and variously assigned to the Great Oolite Series (Middle Jurassic) of the southeast Midlands or to the Kemtish rag facies of the Lower Cretaceous Hythe Beds in the Medway area. These attributions, based on limited scientific evidence, were not persuasive and therefore a comprehensive geological analysis of the whetstones was attempted using petrographic thin-sections, charcoal analysis, and heavy-mineral analysis. This work showed that the whetstones had been made from sandstones in the earliest Cretaceous Weald Clay Formation, at an as-yet unknown site somewhere in the northwest Weald. These rocks are typified by angular, very fine- to fine-grained quartz sand set with shell debris (especially ostracod /pelecypod and echinoderm) and occasional glauconite, chert and feldspar in lustre-mottled calcite cement (Fig. 3.2C). Almost invariably present is wildfire-related, vegetable charcoal exhibiting, especially under the scanning-electron microscope, anatomical structures. The heavy-mineral assemblage (Ch.2) is dominated by zircon and tourmaline, but with small amounts of garnet, staurolite, kyanite and sillimanite, suggesting a partly metamorphic provenance.

As the result of direct examination, or reference to reliable published reports, in which they generally masquerade as Kentish rag, bar-shaped whetstones of Wealden provenance (Fig. 10.3A) have now been identified at more than 40 sites in Roman Britain (Fig. 10.3B). The whetstones cover most of the Roman period, reach as far north as the Scottish Borders, and are recorded on the near-Continent. Apparently marketed from an agency in London, they record a well-organized, major, Roman extractive industry making a high-end product. These well-formed whetstones were sufficiently highly valued as to occasionally serve as votive objects.

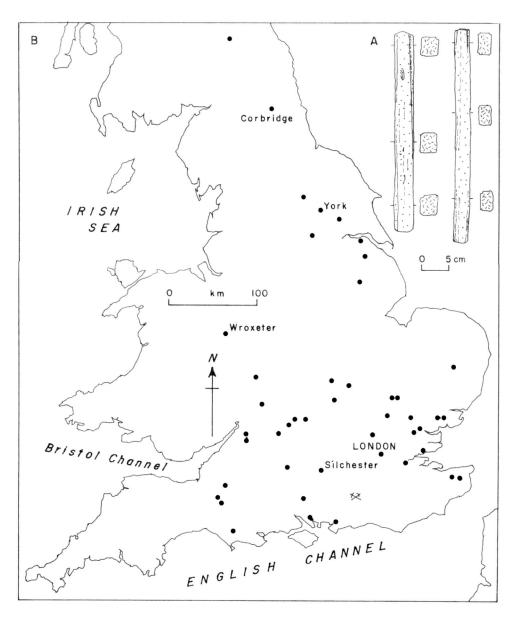

Fig. 10.3 Whetstones of sandstone from the Weald Clay Formation.
A – two complete and little-worn examples. B – distribution in Roman Britain.

Case 3: Iron Age-Roman rotary quern and millstone industries in southern England

The products of three broadly contemporaneous quern industries in southern England have been the subject of careful distributional analysis using petrographic criteria.

The Quartz Conglomerate, in the late Devonian Upper Old Red Sandstone, is a thin unit that crops out extensively in the Forest of Dean-Chepstow area west of the Severn. The rock is a pale grey to pale brownish pink, tough, silica-cemented, medium- to very coarse-grained sandstone with scattered to very abundant, well-rounded pebbles of vein-quartz. Querns (Fig. 10.4A) were produced from the later Iron Age through the Roman Period. They could have been made almost anywhere on the outcrop, but the most likely sources are hillside quarries at Pennalt above the Wye south of Monmouth, where production continued into modern times. Their distribution is unexpected, for they occur mainly east rather than west of the Severn (Fig. 10.5A), and are concentrated in numbers and findspots around Ashton Keynes near Roman Cirencester. Few reached further east than the longitude of the Isle of Wight. The querns could have been floated down the Wye and then carted overland to an agent at Ashton Keynes.

Lodsworth stone is a facies of the early Cretaceous Hythe Beds that crop out in the southern Weald. The rock is a tough, greenish-grey, medium-grained, siliceous sandstone with distinctive patches and whisps of dark-grey, purer chert (Fig. 104B). Rotary querns were produced in great numbers from mid/late Iron Age times through the Roman Period at a quarry at Lodsworth near Midhurst. From there they achieved a main distribution narrowly confined between the longitudes of the Hampshire Avon and the Arun-Wey-Colne, but reached northward into the southeast Midlands (Fig. 10.5B). In contrast to Quartz Conglomerate querns, they seem only to have been marketed from the production site.

Hertfordshire puddingstone, a Pleistocene silcrete, is composed of well-rounded flint pebbles in a non-porous matrix of sililca-cemented quartz sandstone (Fig 10.4C). The rock is massive, tough, and hard to fashion, varying in colour from off-white through grey to orange-brown and even black. It was worked for rotary querns at least during the first and second centuries AD, certainly at a quarry near Braughing in east Hertfordshire and possibly also at Redbourn to the west. The distinctive beehive-shaped querns (Fig. 10.4C) are found mainly east of the Colne and north of the Thames and far into the northeast Midlands (Fig.10.5C).

East Sussex and Kent are largely free of Lodsworth and puddingstone querns. Some other lithology, awaiting identification and systematic analysis, may have been exploited here.

What seems striking and demanding of explanation about the distribution of these artefacts is the northward spread of each lithology while apparently confined between north-south limits that appear largely to exclude potential competitors (Fig. 10.5). These limits transgress tribal boundaries and seem instead to be topographical, identifiable with estuaries and river systems, but not necessarily determined by these features as potential physical *barriers* to trade. It is interesting to note that the distribution of true millstones of Millstone Grit (Upper Carboniferous) has a sharp southerly limit – broadly coincident with the Thames Valley – impinging only slightly on the northward distribution of querns of the three sources discussed. As milling was so vital to civilian and military daily life, it may have been the case that, after deliberately expanding existing native quern/millstone industries, the Roman authorities partitioned southeast Britain into distinct trading areas under separate agents, rather than rely on haphazard production and marketing.

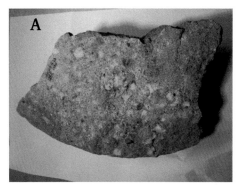

Fig. 10.4 Iron Age-Roman rotary querns A – Quartz Conglomerate (Upper Old Red Sandstone), photo: R. Shaffrey. B - Lodsworth stone, photo: R, Shaffrey. C -Hertfordshire puddingstone, photo: C. Green

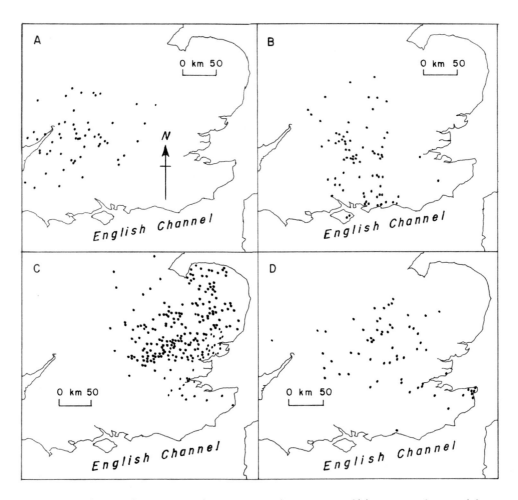

Fig. 10.5 Distribution in late Iron Age and Roman Britain of rotary querns of (A) Quartz Conglomerate (after Shaffrey 2006), (B) Lodsworth stone (after Shaffrey & Roe 2011), (C) Hertfordshire puddingstone (after Green 2011), and (D) millstones of Millstone Grit (after Shaffrey 2015). Note the 'empty' quarter of East Sussex-Kent.

11. Pottery and Brick

Clay formations and their distribution

Pottery for storage, industry, cooking, decoration and display, and brick and tile, a secondary building material, are utilitarian items fashioned from plastic clay rendered strong and durable by subsequently being fired. Here is a crucial pyrotechnology, practised since the Neolithic.

What is required in the parent material is a proportion of clay minerals sufficient, when water is added, to allow a plastic mixture to be made that retains an imposed shape. As place and field names, parish and census records, and patronyms tell, suitable parent geological materials abound in sufficient amounts in most parts of Britain and Ireland. They range from Holocene river and tidal alluvium, through glacial lake muds, to outcropping mudrock formations, and to deeply weathered muddy limestones, granite and gabbro. It is advantageous if, in addition to illite and kaolinite, a good proportion of mixed-layer clay is present (Ch. 2). Clays being heavy, bulky, and costly to transport, early potters and brick-makers mainly relied on local resources. Some clays were of sufficient quality for potting as to be worth importing, for example, fine-grained glacial lake clays from north Devon into South Wales in the early modern period, and china clay from Cornwall to Staffordshire in modern times. Over the last few centuries, however, commercial potters and brick-makers have exploited a few particular formations. These include the Upper Carboniferous mudrocks of British coalfields (e.g. the Etruria Marl), the Triassic Keuper Marl of the Midlands, the Jurassic Oxford Clay and Kimmeridge Clay of the East and South Midlands and Yorkshire, the Eocene Reading Beds and London Clay of the Thames Basin, and the Holocene saltmarsh muds of the Thames, Severn, Humber and other estuaries. Some of these beds, notably the Jurassic mudrocks, have a particular commercial appeal, for they hold sufficient organic matter to be at least partly self-firing. These various mudrocks are found chiefly in 'Young' Britain and the younger, least deformed parts of the 'Old' terrain (Fig. 6.2).

Potting in Britain and Ireland began as a household activity sufficient to meet immediate domestic needs. It developed over the centuries into the main employment of certain settlements, and by Roman times in Britain had become semi-industrialized, with major, exporting potteries springing up in coastal Dorset, the New Forest, Oxfordshire and the East Midlands, amongst other places. The full industrialization and consolidation of the industry is a feature of the early modern and modern periods, as witness the Staffordshire Potteries, which used imported Cornish as well as local clays. The Roman military pioneered in Britain the manufacture of bricks and tiles and their use in building. The craft was revived in late medieval and early modern times. Most parishes had a brickyard or two even up to the middle of the nineteenth century, and many substantial buildings, such as country houses, were erected using bricks burnt on site in temporary yards. In modern times production shifted from small, local yards to a few major centres, such as Bedford (Oxford Clay) and Peterborough (Kimmeridge Clay), exporting bricks and tiles in huge quantities throughout the country.

Clay preparation and forming

Before clay can be made into pots or bricks, it must be dug, cleaned of deleterious matter such as soil, roots or pebbles, and generally mixed thoroughly with water to an even consistency. The stiff, adhesive properties of clay mean that digging it is not always as easy as it sounds; not for nothing were the Holocene, buttery saltmarsh muds of the Severn Estuary called 'bungum'. Most clays, once dug, benefit from being left to *weather* in the open for some months, preferably during the winter, when rain and frost can loosen the mass. Final mixing was done in early times by simply treading water into clay spread over a level floor until the desired consistency and evenness was reached. Mixing is now done mechanically on a large scale in pug mills that chop and homogenize the materials, much as a kitchen blender reduces vegetables.

If the clay was intended for potting, a coarse *temper* was commonly added at the mixing stage, should no natural temper be already present (e.g. rotted granite/gabbro). Tempers could be of almost any material conveniently to hand: quartz sand or fine gravel, crushed rock such as chalk or limestone, broken food shells (e.g. cockles), crushed pottery wasters (grog), slag, and even chopped reeds or straw. The purpose of temper was to reduce shrinkage when pots dried and to improve the resistance of pottery to heating. The analysis of tempers using geological techniques – simply treating pots as rocks – can furnish critical evidence as to the provenance and trading of pottery.

Pottery can be shaped in several ways. At an increasing level of technical sophistication, these are (1) *hand-forming* ('pinching'), (2) *coiling* or *ring-building*, (3) *paddle-and-anvil shaping*, (4) *slab-building*, (5) shaping in a *mould*, (6) *wheel-throwing* (fast/slow), and (7) *slip-casting*. The so-called 'hand-made' pottery found in the early British and Irish archaeological records was produced as pinch-pots or by coiling/ring-building. By Roman times wheel-throwing was well established, and moulding was also used, but some pottery was still at least partly hand-made. Fast wheel-throwing, moulding, and slip-casting are techniques typical of modern mass-production.

Pottery was often decorated prior to firing. Patterns of impressed grooves are common on prehistoric pottery. In Roman times decoration was limited to grooving, rouletting, the addition of a wash of a differently coloured clay, or the application of animal and other figures by squeezing semi-liquid clay from a bag (*en barbotine*). Glazing was extremely rare. Medieval pottery was commonly decorated with applied strips and squibs and coated externally with a green, copper-lead glaze. By early modern times decoration mainly took the form of *sgraffito* designs and trails and patterns in applied clay slips. Glazes had moved mainly to the interiors of pots. Most modern pottery is glazed on both surfaces. There may be a colourful, painted decoration, or exotic/picturesque scenes, applied using transfers, such as the famous willow-pattern.

Traditionally, bricks are made by throwing and pressing a handful or two of plastic clay into hollow, rectangular, brick-shaped, *wooden moulds* with an open top and bottom mounted on a pallet or bench. The surplus clay is scraped off the top and the brick freed from the mould. The resulting bricks preserve on their long, narrow faces characteristic swag-like grooves formed where handfuls of clay came into contact. Early practice, notably Roman and medieval, saw bricks shaped either by hand-moulding or by cutting up large sheets of rolled clay into blocks or slabs of the desired size. Most bricks were made from the middle of the nineteenth century onward either by mechanically *extruding* clay through a die or by *pressing* the partly dried, fragmented material into a mould. Extruded bricks commonly preserve on their larger faces the marks of the wire used to cut them free; faint striae may appear on other faces.. The broken surfaces of moulded bricks often show a 'lumpy' fabric. Bricks made by mechanically chopping up rolled sheets of stiff clay often preserve on their faces series of small, oblique fractures that opened up as the blade sheared through the mass.

Firing

Freshly-made pots and bricks are described as 'green'. They are set out to dry in open-sided sheds before being fired, of which there are many ways. Firing irreversibly converts the clay minerals to a new material known as *ceramic*.

Small amounts of pottery can be fired using any convenient carbon fuel in shallow scoops in the soil but moderate amounts demand the use of open bonfires or clamps built on the ground surface. Bricks can also be fired in clamps. The largest, as seen in India, are huge, migrating, snake-like structures in which fresh bricks and fuel are heaped up and set on fire at one end while sufficiently cooled, burnt bricks are removed at the other. Pottery and bricks were normally fired in batches using kilns (Fig. 11.1), because these allowed more precise control of the kiln atmosphere and firing temperature and led to greater output. They are of many different designs, tending to increase in size over time. Modern ones for bricks and tiles are huge and continuously operated.

Like Jacob's coat, bricks are of many colours: grey, blue-purplish grey, buff, pale brown, orange, bright red, dark red, dull yellow, greenish yellow, black. Some of these colours can also be found in pottery, for example, the orange-red of Roman and medieval 'oxidised' wares, and the grey-black of the 'reduced' wares of these periods. Pottery and bricks are variable in consistency. Some are soft and porous and prone to weathering in the archaeological environment. Others, fired at higher temperatures, are compact, hard and almost rock-like, for example, early modern / modern stoneware pottery and modern blue or dark red engineering bricks. Grey and blue bricks were often crudely glazed by throwing salt at a late stage into the kiln.

Fig. 11.1 The early-modern bottle-kiln at Nettlebed, South Oxfordshire

Several poorly-understood factors control this range of properties. They include the mineralogy and chemistry of the clay and any introduced additives, such as the chalk ('malming') used by some brick-makers. Fabric strength and consistency is largely a function of kiln temperature, high values favouring the sintering of the mineral particles. A low proportion of oxygen in the kiln atmosphere – reducing conditions – produces grey-black shades. An oxidizing environment guarantees orange-red colours.

Building in brick

Brick as a building material is more typical of southeastern and midland Britain than any other part of the country. It is highly versatile, flexible and convenient, because it is cheap, comes in standardized sizes, shapes and colours, and can be sawn and easily carved.

Bricks can be laid, or *bonded*, in very many ways depending on the size and purpose of the structure under erection (Fig. 11.2). These bonds are distinguished by the attitude of the bricks and the way the long (stretcher) and short (header) side faces are arranged. *Stretcher bond* (Fig. 11.2A) has been used over the last 75-100 years to build cavity walls. The *English bond* (Fig. 11.2B), of alternate stretcher and header courses, was often used for large buildings and structures that called for strong walls. *Header bond* is another strong arrangement Fig. 11.3C). In wide use from the seventeenth and eighteenth centuries was fashionable *Flemish Bond* (Fig. 11.4D). *Rat-trap bond* resembles it, except that the bricks are laid on the narrow face rather than on the bed (Fig. 11.2E). To make arches

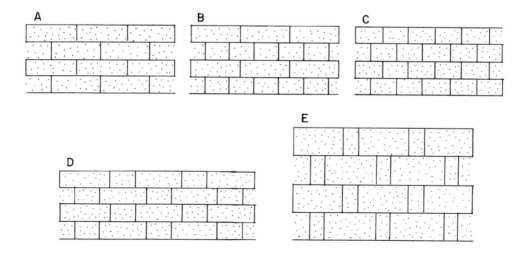

Fig. 11.2 Some common brick bonds (external views). A – stretcher bond. B – English bond. C –header bond D – Flemish bond. E – rat-trap bond

such as over windows or doors, bricks could either be sawn or rubbed to shape or special voussoir-shaped bricks purchased. Fancy chimney pots and finials, as in grand Tudor houses, could be *carved* in situ from bricks already laid. The size, shape and style of bonding of bricks has gradually varied over time, and these changes, calibrated against documentary sources, can be exploited to date historic buildings.

Victorian architects in particular, such as W. Butterfield, G.E. Street, and S.S. Teulon, all participants in the Decorative Movement, were enthusiastic users of *polychrome brickwork*, in which differently-coloured bricks combined to make patterns and designs. Such polychrome brickwork was not just confined to churches, where it accompanied colourful encaustic floor tiles, but appeared widely in the fabric of ordinary houses of the later nineteenth century (Fig. 11.3). Patterned encaustic tiles had been used in medieval times to floor churches and important buldings.

Fig. 11.3 Polychrome brickwork in a nineteenth-century street, Reading, Berkshire

Clay was also used to make fancy red or buff, decorative items – columns, statues, vases, roundels, plaques, finials, chimney pots – as well as building blocks larger than bricks, that generally go by the name of *terracotta*. Allied to these are the beautiful, moulded/carved human and animal figures, finials, vases, urns and plaques produced in the late eighteenth and early nineteenth centuries by Mrs Eleanor Coade's London company in the buff-coloured artificial stone carrying her name. They appear in many fashionable residences and country houses.

Further Reading

Brunskill, R.W. 1990. *Brick building in Britain.* London, Gollancz

Davey, E. & Roseff, R. 2007. *Herefordshire bricks and brickmakers.* Little Logaston, Logaston Peess

Field, R. 2004. *Geometric patterns from tiles and brickwork.* Diss, Tarquin Publications

Grant, A. 1983. *North Devon pottery. The seventeenth century.* Exeter, University of Exeter Press.

Harrad, L. 2004. Gabbroic clay sources in Cornwall: a petrographic study of prehistoric pottery and clay samples. *Oxford Journal of Arechaeology* 43, 271-286

Lucas, R. 2000. Brickmaking on Norfolk commons. *Norfolk Archaeology* 43, 457-468

McNeill, C. 2015. *Potters of Kirkaldy. The home of Wemyss Ware.* Stroud, Amberly Press

Peacock, D.P.S. 1982. *Pottery in the Roman world.* London, Longman

Quinn, P.S. 2013. *Ceramic petrography.* Oxford, Archaeopress

Van Lemmen, H. 2002. *Architectural ceramics.* Princes Risborough, Shire

Williams, D.F. 1977. The Romano-British black-burnished ware industry: an essay in characterization by heavy minerals analysis, in D.P.S.Peacock (ed.) *Pottery and early commerce.* London, Academic Press, 163-220

Case 1: Prehistoric gabbro-tempered pottery, Southwest England

Gabbro (Ch. 3) s sufficiently unusual in southern Britain as to be limited to just one area, the Lizard Peninsula on the south Cornish coast, where it is associated with schists and serpentinites. This basic igneous rock is a readily-altered, coarse-textured mixture of calcium plagioclase, clinopyroxene and olivine, with some magnetite. Gabbro-tempered coarse wares were made on the Lizard Peninsular from Neolithic (chiefly) through to Roman times. At least six distinct gabbroic pottery fabrics have been distinguished by detailed petrographic and geochemical analysis. Complementary work on putative source clays from the Lizard strongly suggests that these represent production at several different locations and times. In a pattern typical of product fall-off, Neolithic gabbro-tempered pottery declines steeply in relative abundance, compared to competitors, eastward from the Lizard Peninsula, in a manner suggesting mainly coastal trading (Fig. 11.4), and also northward from the coast, perhaps along the valleys of the R. Avon and Wylie.

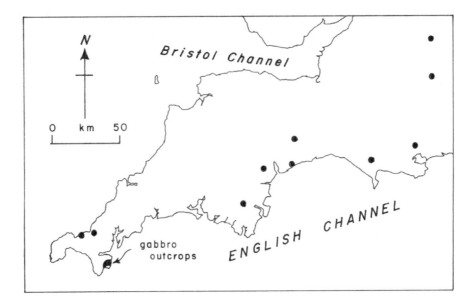

Fig. 11.4 The distribution of Neolithic gabbro=-tempered ware in Southwest England

Case 2: Roman Southeast Dorset Black-burnished Ware

The commonest coarse pottery known from Roman Britain is a black-burnished fabric divided between type I, found almost everywhere by the fourth century AD, and type II, restricted to eastern England and southeast Scotland. Chemical analyses confirmed this division, but left the hotly-debated question of the production sites unresolved. That issue was finally and convincingly settled by means of heavy-mineral analyses (Ch. 2) of the abundant sand temper recovered from the pottery, together with putative temper deposits. True type I pottery – Southeast Dorset Black-Burnished Ware No. I – is typified by a tourmaline-rich heavy-mineral assemblage, consistent with production in the Wareham-Poole Harbour area of Dorset. Here tourmaline derived from the Cornubian granites to the west is common in the Tertiary sands, and numerous Romano-British bonfire/clamp firing sites are known. Black-burnished wares are a critical dating tool, because of their wide distribution (including the near-Continent), abundance, and long time-range.

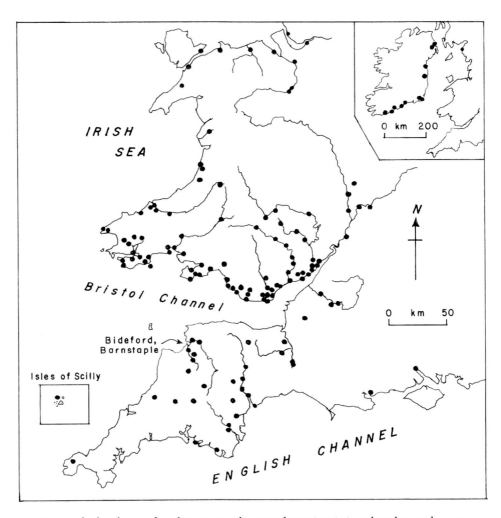

Fig. 11.5 The distribution of North Devon Gravel-tempered Ware in Britain and southern and eastern Ireland, based on archaeological and documentary sources

Case 3: North Devon Gravel-Tempered Ware

The north Devon towns of Bideford on the Torridge estuary, Great Torrington on the freshwater Torridge, and Barnstaple on the Taw estuary once hosted one of Britain's greatest pottery industries prior to the emergence of the Staffordshire enterprises. Begun in the early seventeenth century, the industry lasted into the early eighteenth, when it stagnated and slowly died. The industry produced Gravel-Tempered Ware, a wide range of glazed and often decorated dishes, bowls, jugs, pipkins, porringers, chafing dishes, chamber pots, tankards, jars, cups, commemorative items, candlesticks, dripping pans, firedogs and, especially popular in southeast Wales, baking ovens. The work was controlled by several, partly-interconnected Devon families, which included merchants and traders.

Arrangements for the supply of raw materials and dispersal of the finished products were fairly complicated. Stone-free, very fine-grained, glacial-lake clay was shipped to Bideford (8 km) and Barnstaple (5 km) from the village of Fremington, where it was dug. Slip for piped or *sgraffito* decoration was made using a white, kaolin-rich ball-clay of Miocene date from the village of Peters Marland 13 km south of Bideford. Gravel (quartz, feldspar, mica) for temper was brought by boat down the rivers, but it was also dug from the estuaries, as petrographic work clearly reveals the presence in some pots of shell debris and modern foraminifera. Galena (PbS) for glazing came from nearby mines but also from Mendip and west Wales. Coal, the main fuel, came chiefly from Swansea and Pembrokeshire, across the Bristol Channel. The glazes were flash-liquefied by burning in the kilns furze (gorse), gathered from local moorlands. The pottery was traded by sea to coastal sites in Southwest England, Wales, and the Severn Estuary, and thence inland, and also to Cumbria and the south and east coasts of Ireland (Fig. 11.5). There was some overland trading in Devon and Cornwall. Of especial note is the considerable trade across the Atlantic with the developing colonies in New England, Maryland, Virginia and the West Indies. It was here that the pottery first received serious archaeological attention.

12. Metallurgical Landscapes

Metals

To the archaeologist, metals are heavy, dense, opaque, fusible and malleable materials that can be given a useful or decorative shape by casting, forging, cutting, hammering or filing. Gold, silver, and copper are the few metals found native in Britain and Ireland, albeit very rarely. Elsewhere meteoritic nickel-iron was known and exploited. These and other metals had in the main to be produced from natural ores by the pyrotechnology of smelting. The discovery of metals was either accidental – the unexpected effects of camp or wild fires – or the result of deliberate, alchemist-like experimentation with rocks of unusual colour or weightiness. The exploitation of metallic ores led to the creation of complex and often extensive *metallurgical landscapes*, in which mines, smelters, forges, carbon fuel sources, and means of transport were all essential, interconnected parts. Less tangible and discoverable, but just as important, were the various interactions between early metallurgists and the community at large.

Of the precious metals, *gold* (Au) was used for decoration, jewellery, tableware, and coinage. It was valued for its ponderous quality (specific gravity, SG=19.3), ease of working, failure to corrode, and bright colour, resembling muted sunlight. *Silver* (Ag, SG=*c.* 11)), another precious metal, is a bright, white substance but susceptible to corrosion and only half as ponderous as gold. It was avidly sought chiefly for coinage, especially in Roman and again in medieval-early modern times, but also for tableware and less prestigious ornaments.

The base metals known in the ancient world are in the main less ponderous than the precious ones and all more or less subject to corrosion. *Copper* (Cu, SG=8.9) is red, soft and malleable, with a melting point of *c.* 1100°C. It could therefore be cast as well as worked in solid form. Two other metals, tin and zinc, were commonly alloyed with copper. *Tin* (Sn, SG=7.3) is a bright, white metal with the low melting point of 232°C. Alloyed with copper it forms yellow-brown *bronze*. Bronze had many uses, from the making of all kinds of domestic and personal items to coinage, axes and swords, and cannons. *Zinc* (Zn, SG=7.2) is a difficult metal and in Britan and Ireland may never have been made as such. It melts at 419°C but immediately forms a vapour from which it must be condensed. Blended with copper it forms yellow *brasses*, made in Roman and later times by heating together granulated copper and a zinc ore. The much-valued, soft, grey metal *lead* (Pb, SG=11.3) was readily made into sheets, tubes, pipes, and containers. Lead was alloyed with tin in Roman and later Britain, to give *pewter*, used for dishes, plates, drinking vessels, and ornaments. The only metal liquid at room temperature is *mercury* (Hg, SG=13), used as an alloy with gold for gilding items of base metal.

Iron (Fe, SG=*c.* 7.5), the second most abundant element in the Earth's crust, was of crucial importance for tools and weaponry from the Iron Age onwards. Its ores are widespread, many and various, although the high melting point of 1538°C meant that liquid iron was obtainable only when the blast furnace appeared in late medieval times. It forms various *steels* when combined with carbon and metals such as chromium, manganese, nickel, tungsten, and vanadium.

Metalliferous ores and their genesis

A host of minerals include a useful metal in their composition, but few are plentiful or rich enough to have been worth exploiting. Chemically, the metalliferous ores valued in Britain and Ireland, and likely to be found archaeologically, are of five main kinds: hydrated oxides, oxides, carbonates, sulphides and silicates. These represent a combination of the metal with water/hydroxyl radical, oxygen, carbonic acid, sulphur or silica with oxygen as the case may be. Gold occurs native in Britain and Ireland. The refining of argentiferous lead by cupellation produced most silver.

The chief hydrated oxide ores are those of iron, namely, the yellow ochres *limonite* (brown haematite) and *bog iron ore* ($2Fe_2O_3.3H_2O$)), and *goethite* (Fe(OH)), also used as pigments.

Oxide ores are very important. Copper has *cuprite* (Cu_2O) and tin *cassiterite* (SnO_2). Those of iron are *magnetite* (Fe_3O_4) and red *haematite* (Fe_2O_3), the latter also used as a pigment when ochreous (reddle).

Several metals occur as carbonates. Copper forms the hydrated carbonates *malachite* ($CuCO_3.Cu(OH)_2$) and *azurite* ($2CuCO_3.Cu(OH)_2$), lead *cerrusite* ($PbCO_3$), and iron *siderite* ($FeCO_3$).

Sulphide ores were especially valued. Copper has *chalcocite* (Cu_2S) and the more plentiful double sulphide *chalcopyrite* ($CuFeS_2$), lead *galena* (PbS), zinc *sphalerite* (ZnS), and iron *pyrites/marcasite* (FeS_2), and sulphur-enriched *pyrrhotite* (magnetic pyrites). Mercury occurs as the red sulphide *cinnabar* (HgS), albeit extremely rarely in Britain.

Chamosite is a hydrated iron-aluminium silicate and an important sedimentary ore of iron. Of some use as a copper ore is the hydrated silicate *chrysocolla* ($CuSiO_3.2H_2O$).

At a global level, ore genesis is a vast and highly specialized subject. So far as British and Irish resources are concerned, however, only a few geological processes need be noted. Hot chemically reactive gases are released into the surrounding rock during the later stages of the consolidation of intrusive magmas. These give rise to *pneumatolytic mineral veins* likely to include tin, iron and copper ores. Hot chemically reactive waters carrying dissolved minerals also form late during magma consolidation, especially where granites are concerned. The resulting *hydrothermal deposits*, found as veins in the chilled magma and as infilled joints and fissures in the surrounding country rock, are commonly rich in copper, lead and zinc ores. The minerals may even fill the pores of sandstones where these are present in the surrounding rock. *Metasomatic ore deposits* result from the partial replacement by an ore mineral of a rock such as limestone. The goethite and haematite ores of the Forest of Dean and the southern Lake District, and the siderite replacements of limestones in the Cheviot Hills, are of hydrothermal-metasomatic origin. The former probably arose when iron-enriched waters from the Trias circulated in voids and opened joints in the Lower Carboniferous Limestone below, which had experienced some karstification. Although generally of low grade, *sedimentary ore deposits* are valued because of their comparative abundance and ease of working. The early diagenesis of organic-rich muds gives rise as mudrocks accumulate to nodules and bands of clay-ironstone, a mixture of clay minerals and siderite. Clay-ironstones are widely represented in the shales of the Upper Carboniferous Coal Measures, where they conveniently occur in association with coals (e.g. South Wales, Coalbrookdale in Shropshire, Scottish Midlands), and in related strata such as the Lower Cretaceous Weald Clay Formation. Of more restricted occurrence are bedded chamosite oolite ores found in the Jurassic of the East Midlands. *Placer deposits* are enrichments due to hydraulic sorting by particle size, shape and density in streams or on wave-swept beaches. Placer cassiterite was worked in Southwest England, and placer gold in Scotland, Wales and Ireland.

Outcropping ore bodies experience weathering for the same reasons as rocks generally. New minerals may form, chiefly through oxidation and hydration, and the descending waters may secondarily enrich ores lower down in the ore mass.

Ore fields and mining

It is true only up to a point that 'gold is where you find it,' for the distribution of metalliferous deposits (Fig. 12.1) is under strict and discoverable geological controls. Especially important is the presence of concealed to partly exposed granite batholiths, as in the Scottish Southern Uplands, Cumbria, the northern and southern Pennines, and Southwest England.

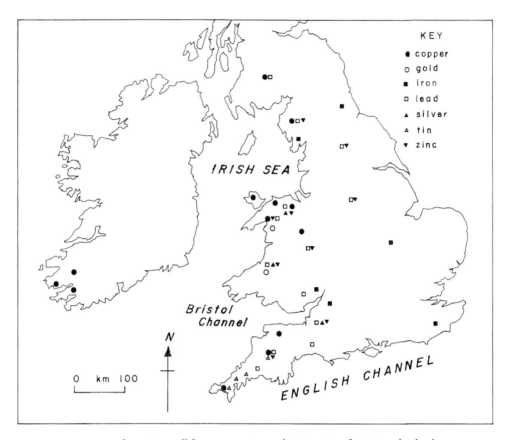

KEY

● copper
○ gold
■ iron
□ lead
▲ silver
▲ tin
▼ zinc

Fig. 12.1 The main metalliferous ore regions and mining sites of Britain and Ireland

From the standpoint of mineralization, Britain and Ireland can also be divided between the 'Old' and the 'Young' terrains introduced in Ch. 6.

In Ireland mineralization is mainly associated with Devonian and Carboniferous strata and especially with the intensely folded rocks of these dates in the southwest. Here occur numerous sites, including that of Mount Gabriel, where sulphide and sedimentary ores associated especially with the Old Red Sandstone were worked for copper in the Bronze Age. Vein and placer gold are also known from Ireland.

Britain is more complex. The 'Old' terrain in the north and west comprises Scotland, the Pennines and Cumbria, Wales, and Southwest England, where variously deformed pre-Palaeozoic, Lower Palaeozoic, and earlier Upper Palaeozoic (Devonian-Carboniferous) rocks predominate (Table 4.1). To the east and southeast, and flanking the Pennines, is 'Young' terrain. Here predominate little-deformed later Upper Palaeozoic (Permo-Triassic), Mesozoic and Cenozoic sediments.

Non-ferrous mineralization is widespread and locally plentiful in Old Britain. Lead, copper and gold ores occur in Southern Scotland. Deposits of lead and zinc, and some of copper, are found in Cumbria, the north and south Pennines, Wales and the Welsh Borders, the Mendips, and Southwest England. The chief Bronze Age copper-mining areas were at Parys Mountain (Anglesey) and the Great Orme in North Wales, the Llancynfelin-Cwmystwyth district of Central Wales, and Alderley Edge-Mottram in the northwest Midlands. Dolaucothi in southwest Wales was an important Roman source of gold. From prehistoric times onward Southwest England yielded much vein and placer tin.

Iron ores abound in both Young and Old Britain. Goethite-haematite deposits associated with the Lower Carboniferous Limestone have been worked in the Forest of Dean and north Mendips from the Iron Age to modern times. A similar orefield occurs in south Cumbria. Especially in early-modern times, clay-ironstone was exploited in almost every outcrop of the Upper Carboniferous Coal Measures, in Scotland, on the flanks of the Pennines, in the West Midlands, and in South Wales. Earliest Cretaceous clay-ironstone from the Weald of Kent was an important resource in the Roman and medieval periods. Various sedimentary ores of the Jurassic ridge in the East Midlands were exploited during Roman and again in modern times.

Geological circumstances largely determine how an ore-body is mined: its composition, size, shape and attitude, and the character of the enclosing sediments or rocks. In Roman times the gently dipping Jurassic beds of the East Midlands, and the early Cretaceous clays of the Weald, were worked in scattered *opencast pits* on the outcrop. There is as yet no firm proof of working from bell-pits or from hillside adits. The ores of the Forest of Dean were worked in deep cuttings open to the sky, known as scowles, that followed the narrow deposits. A line of pits might represent an ore body that varied in quality along its length. *Stoping* was the chief method of working steeply inclined ore veins, such as in the south Pennines and Southwest England. As a narrow vein is excavated downward from an open rake, braced timber working platforms with connecting ladders are erected within the growing space. Soft strata could be excavated using picks and levers, but hard rock, prior to the late medieval introduction of gunpowder, was removed by *fire-setting*. A wood fire was built against the face to be advanced and the hot rock then fractured by dousing with cold water. By pounding the face with stone hammers the shattered rock with its ore was further weakened and the pieces finally levered off. Many prehistoric mines have yielded hundreds of grooved cobbles that had served as hammers when lashed to handles.

Hushing, a technique of hydraulic mining known from the Bronze Age, was practised at Roman and later gold, copper and lead mines in Britain. As surviving earthworks and surface scars reveal, the process involved the release of a torrent of water from an artificial reservoir, in order to flush loose overburden from the slope covering an ore body.

Smelting and refining

The ore sent from a mine generally requires further treatment before smelting. Ore fragments need to be picked out, the larger ones crushed to an appropriate size, and the whole washed and graded Non-oxide ores demand chemical treatment prior to smelting. Sulphide ores were roasted in pits or clamps with a carbon fuel in order to remove sulphur, leaving behind the smeltable metal oxide. Similarly, by calcining carbonate ores carbon dioxide could be driven off. Some hydrated and silicate ores benefitted from pre-heating.

Chemically, the process of smelting is one of *reduction*, in which the metal oxide (MO) is converted to the metal (M) by reacting at a high temperature with carbon monoxide (CO) produced by the incomplete burning of an accompanying carbon fuel:

$$MO + CO = M + CO_2$$

carbon dioxide escaping into the atmosphere. Up until late early modern times, when coke made by the dry-distillation of coal was introduced, charcoal was invariably the favoured carbon fuel. In practice the smelting process was often more complicated. In the case of copper, commonly made from chalcopyrite, the iron had first to be removed as a slag at the roasting stage before the copper oxide remaining could be finally smelted. The process for iron was also not straightforward prior to the introduction of the blast furnace in latest medieval times, as furnace temperatures did not reach its melting point. The first stage in the *bloomery process* was the making of a solid bloom of iron contaminated with slag. The second stage saw the contaminants expelled by hot-forging.

Before the introduction of the blast furnace, iron-smelting was a batch process that yielded comparatively little metal at a time. Several kinds of early bloomery furnace are known, ranging from simple scoops in the soil to slag-pit furnaces and re-usable shaft furnaces from which tell-tale, fluid slag could be tapped as a lava-like stream (Fig. 12.2). A system of bellows was used to force air through tuyeres into the furnace charge. Iron production increased by orders of magnitude with the introduction of the continuous *blast furnace*, run

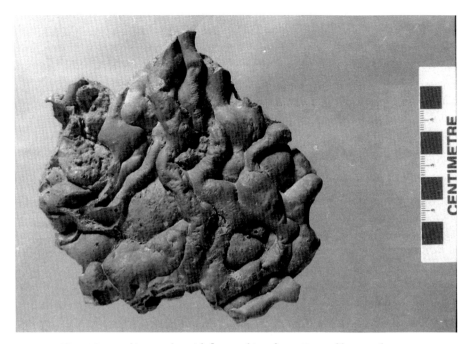

Fig.12.2 Iron-making tap slag with flow markings from a Roman bloomery furnace

for months, as in Coalbrookdale (Shropshire), the Forest of Dean, and South Wales, from which both liquid metal and slag could be repeatedly withdrawn. The liquid iron could at once be cast into useful shapes, and to make wrought iron and steels the metal was 'puddled' in a reverberatory furnace. The latter process was abandoned in the mid nineteenth century when Henry Bessemer introduced his 'converter' for steel-making. The latter burned the carbon out of the raw iron in a controlled way, and also allowed small amounts of other metals to be added to form alloys.

The insatiable demand for silver coinage was largely met by refining argentiferous lead, such as was mined in the Mendips, Somersest, and the Tamar Valley, Devon. During *cupellation* air is blown over molten, argentiferous lead and the solid lead oxide repeatedly skimmed off until only silver remains. The lead is regained by smelting the oxide in the normal way.

Smelting and refining give rise to a rich variety of residues that survive well in archaeological environments. Iron-making slags vary from basin-shaped masses from simple bloomery furnaces to the liquid slags produced by shaft furnaces and blast furnaces. Flowing slags were also produced during copper-smelting, but these can be distinguished from iron-making slags, which they resemble in hand-specimen, by chemical analysis. The use of chemical methods is in fact essential when attempting to characertize slags and ores.

Further Reading

Bowden, M. 2000. *Furness iron*, Swindon, English Heritage

Burnham, B. & Burnham, H. 2004. *Dolaucothi-Pumpsaint. Survey and excavation at a Roman gold-mining complex 1987-1999*. Oxford, Oxbow Books

Cooper, M.P. & Stanley, C.J. 1990. *Minerals of the English Lake District: Caldbeck Fells*. London, Natural History Museum.

Gribble, C.D, 1988. *Rutley's Elements of Mineralogy*, 27th ed., Berlin, Springer

Hughes, S. 2000. *Copperopolis. Landscapes of the early industrial period in Swansea*. Aberystwyth, Royal Commission on Ancient and Historical Monuments of Wales

O'Brien, W. 1996. *Bronze Age copper mining in Britain and Ireland*. Princes Risborough, Shire

Rippon, S., Claughton, P., & Smart. C. 2009. *Mining in a medieval landscape. The Royal Silver Mines of the Tamar Valley*. Exeter, Exeter University Press

Schrüfer-Kolb, I. 2004. *Roman iron production in Britain*. Oxford, Archaeopress

Todd, M. 2007. *Roman mining in Somerset. Excavations at Charterhouse on Mendip 1993-1995*. Exeter, The Mint Press

Tylecote, R.F. 1987. *The early history of metallurgy in Europe*. Harlow, Longman

---

Case 1: Bronze Age copper-mining at Mount Gabriel, southwest Ireland

Mount Gabriel is perhaps the most important of the many sites in the mountainous far southwest of Ireland at which Bronze Age people mined for copper. The ores worked here are stratiform disseminations of malachite and some sulphides in sandstones and conglomerates interleaved with mudrocks in the intensely-folded and cleaved Old Red Sandstone of the area. Because the ore bodies are discontinuous, the mines are in the form of numerous, opencast pits and a few short adits scattered over the steep hillsides, rather than deep, lengthy workings. Cobble hammers, a twisted withy handle, a wooden shovel, wooden levers, and partially-burnt pine splint-lights testify to the manner of working, which almost certainly began with fire-setting. The piles of broken and crushed rock suggest that some kind of ore preparation took place on-site. No smelting sites are known at Mount Gabriel, but finds of Bronze Age metal artefacts decline in frequency with increasing distance from the locality. The mines flourished in the second millennium BC.

Case 2: Roman gold-mining at Dolaucothi, southwest Wales

An unusually complicated metallurgical landscape is found at the Roman gold-mining complex of Dolauucothi, between Llandeilo and Lampeter, southwest Wales (Fig. 12.3). The site lies at an altitude of *c.* 150-250 m on the steep southeastern side of the valley of the Afon Cothi. Here gold occurs as disseminations of the native metal in quartz veins and lodes that cut steeply dipping early Silurian slaty shales, and perhaps also as local placers. Roman working probably began around AD 70, initially under legionary control, and continued for over 200 years. The dating of the many features, however, is not as secure as might be wished, and there is a strong overprint of modern activities.

Roman mining (Fig. 12.4) was undertaken in narrow rakes, large deep opencast pits, lengthy adits, and also stopes, one of which preserved a water-wheel, part of a drainage system. Water supplies were vital for both mining and processing. Two substantial, main leat systems have been identified. The earlier, the Annell Leat, is some 7 km long and feeds a large tank upslope from one of the biggest opencast pits. The Cothi Leat is later and taps the river *c.* 11 km upstream to the northeast. Short sections of other leats are known, each feeding a substantial tank. These are clay-lined and cut back into bedrock. They probably served a number of purposes, including hushing and the processing on washing tables of finely crushed and ground ore to separate the gold particles. A water-mill may have been present. A Roman fort, a point of control, and road are known at Pumpsaint, a little over a kilometre to the west.

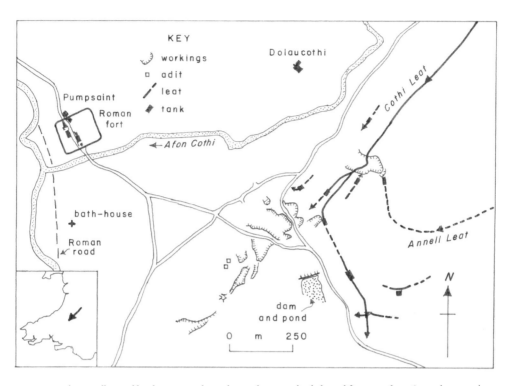

Fig. 12.3 The metallurgical landscape at Dolaucothi, southwest Wales (adapted from Burnham & Burnham 2004)

Fig. 12.4 A mine entrance at Dolaucothi, southwest Wales. (Photo: David Ross, www.britainexpress.com)

Case 3: Roman lead-mining at Charterhouse-on-Mendip, Somerset

The site lies at an altitude of *c.* 300 m on the high plateau of the South Mendips. Mining has occurred intermittently here from certainly the Roman period up to the nineteenth century. The discovery in the area of lead ingots dated to AD 49 suggests that the Romans began to exploit the economic potential of the site almost immediately upon the Claudian invasion of AD 43, establishing fortlets and a small town (name unknown) with an amphitheatre. The chief ore mined was argentiferous galena associated with some haematite occurring in steep veins and lodes in the Lower Carboniferous Limestone. The mines were shallow rakes open to the sky, dug without evidence for fire-setting. Finds at the site include galena ore, lead slag, and smelting debris. The production of silver by refining the lead in cupels seem to have taken place, for some early ingots carry inscriptions declaring them to be silver-free. Mining began under legionary control (*Legio II Augusta*) but by *c.* AD 60 was managed by contractors. Some production persisted into the third and fourth centuries.

Case 4: Early modern-modern iron-making in southern Cumbria

This industry is sufficiently recent that the metallurgical landscape it created is well-preserved and open to investigation in its full complexity, using both archaeological and documentary evidence (Fig. 12.5). The ore chiefly worked, in open pits and later deep mines, was metasomatic haematite (kidney ore) found in the Lower Carboniferous Limestone Series of the rugged Furness peninsula in the extreme south of the Cumbrian massif. There were a few mines in Eskdale to the northwest. As in the Roman Severn Vale, exploiting Forest of Dean ores, smelting and processing sites typically lay far removed from the mines. Large, charcoal-fired, masonry blast furnaces were chiefly used, sited on streams with an assured, strong, year-round flow to drive mills that powered the blast. Several bloomeries were also operated during the earlier history of the industry, the most northerly occurring in the Langdale area in the central massif. Five substantial woodlands were managed for seasonal charcoal production, as testified by the survival within them of burning platforms (pitsteads) and worker's huts. A complex pattern of transport routes – by estuary, river, lake (Coniston, Windermere), track, and road – was developed to bring to the smelters ore and limestone flux from the mines and quarries and charcoal from the woods, and to shift to market in reverse the iron produced.

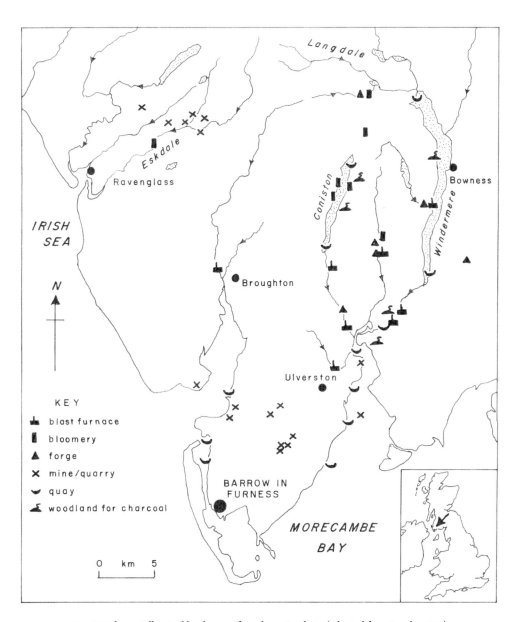

Fig. 12.5 The metallurgical landscape of southern Cumbria, (adapted from Bowden 2000)

Glossary

active layer: the uppermost part of a body of permafrost that thaws and refreezes annually

adit: a horizontal or gently sloping entrance to, or passage within, a mine

axis of symmetry: an imaginary axis, rotation about which causes the faces of a crystal to reappear in the same position

bedding: the different kinds of layering shown by especially sedimentary deposits, ranging in scale from a few metres down to a millimetre or so

BP: Before Present (years before 1950), the standard way of reporting scientific dates by radiocarbon analysis of archaeological materials

Coade stone: a tough, yellowish, artificial stone, really a ceramic, made from pipe- clay, flint, sand, glass and already-fired stoneware, all ground to a very fine powder

corbelling: a technique for making an arch or roof by laying thin slabs of stone in succession so that later extend inward beyond earlier

crag-and-tail: a large glacial landform elongated parallel with the direction of ice-flow, consisting of a rock outcrop at the head and a sheltered accumulation of till or other sediment to leeward. The most famous is that on which the Royal Mile stands between the Castle and the Palace of Holyrood in Edinburgh

desiccation cracks: patterns of vertical, polygonal fractures that result when mud dries out in the sun and wind

diagenesis: a group of physical and chemical geological processes, including compaction, cementation, and anaeorobic organic-matter decay, that act on sediments soon after deposition. New minerals and textures can result

duricrusts: a range of lithified materials that result from superficial processes, including calcretes, ferricretes and silcretes

eddies (of turbulence): a quantity of a moving fluid, such as in a river, the wind, or a tidal stream, the constituent elements of which, while experiencing translation, are in strong, random, three-dimensional motion

gangue: the unwanted minerals recovered along with metallic ores during mining

half-life: the time required for a radioactive isotope to lose half its mass

halophytes: in Britain and Ireland, a large and diverse group of plants able to tolerate to varying degrees according to species the severe conditions that prevail on coastal marshes. On a typical marsh the plants are zoned vertically within the tidal range according to their degree of tolerance, from so-called pioneer species, found at the boundary with a mudflat, to forms that flourish only around the level of high water

hydration: the incorporation of water (H_2O) into the composition of a chemical compound

hydraulic sorting: the action of currents on mixtures of mineral particles is to sort them into like-with-like fractions by size, shape and density. Under wave-action on beaches, for example, coloured heavy minerals often become sorted into distinct laminae, visible on the beach surface or by carefully digging beneath. A gold placer is a product of such sorting

hydroxyl radical: the grouping (OH), behaving chemically like an element

ice-wedge cast: the cast in some kind of sediment, often wind-blown sand, of wedge- shaped bodies of ice that formed in the active layer and uppermost permafrost during the coldest conditions in the periglacial zone. They commonly form polygonal patterns on a 10-metre scale and have been known to mislead archaeologists

involutions: swirling, distorted patterns of bedding arising from freeze-thaw in the sediments of a periglacial active layer

isomorphous series: a group of chemically related minerals that all share the same or closely similar crystal form

joints: the natural fractures that divide up rocks, in sedimentary rocks normally perpendicular to bedding and increasing in spacing with bed thickness

karstification: the process of creating fissures, caves and a karst landscape in a limestone area as the result of the solution of the rock by ground and subterranean water

kettle holes: substantial depressions now occupied by ponds or meres where substantial bodies of ice became trapped and stagnated during deglaciation

kinetic energy: the energy that a body has by reason of its motion

lamination: the finest element of the bedding found in sedimentary rocks, especially sandstones. Most laminae are no more than a millimetre or so thick but can extend for many decimetres

leat: a narrow, artificial channel that taps a natural stream and feeds a reservoir, building, water mill, or mineral processing site

Little Ice Age: a period, broadly from the thirteenth to the nineteenth centuries, when cold, dry 'Continental' conditions prevailed in Britain and Ireland

magma: liquid rock at a high temperature, appearing at the Earth's surface as white-hot lava

mass-movements: downslope displacements of soil, sediment or rock masses under the action of gravity. They include violent, catastrophic movements, such as landslides and rock-falls, but also the much slower processes of solifluction and soil-creep. Mass-movement, often on a colossal scale, are also known from lakes and the ocean margins

millstone (excluding donkey/Pompeiian-style mills): in Iron Age and Roman Britain, a flat, circular pieces of stone with a central hole, generally thinner and wider than a typical rotary quern

Moh's Hardness Scale: a sequence of minerals of increasing relative hardness – talc, gypsum, calcite, fluorspar, apatite, orthoclalse feldspar, quartz, topaz, corundum, diamond

oxidation: the combination of oxygen with a metal or compound

patterned ground: generally either linear (sloping areas) or polygonal (level areas) patterns of soil marks seen on a scale of several to many metres in former periglacial zones. In association with ice-wedges and involutions, and common in Britain and Ireland, they are features of the active layer.

permeability: the property of a sediment of allowing the passage of a fluid through its pores; sands are highly permeable, mudrocks of very low permeability

phytoplankton: various algae, the bottom of the food chain, that live in sunlit surface ocean waters, typically forming seasonal blooms

plane of symmetry: an imaginary surface dividing a crystal or figure into two, mirror-image halves

porosity: the property of a rock or sediment of having internal voids, which need not be connected

potential energy: the energy that a body has by reason of its distance from the Earth's centre

pyrotechnology: any process that requires the use of high temperatures, such as lime and gypsum burning, potting, brick and tile making, smelting of metallic ores, glass making, metal refining, forging and smithing

reduction: the removal of oxygen from the composition of a chemical compound

reverberatory furnace: a furnace so shaped that metal or items placed on its wide, flat bed are heated by the stream of hot gas passing overhead

roches moutonées: exposed rock smoothed by ice to look like resting sheep

salts: compounds formed by the reaction of an element or radical with an acid

sgraffito: a design on a pot made by scratching through a thin layer of applied slip of a different colour to an earlier layer of slip or to the body of the pot

silcrete: a duricrust cemented with silica, typically a sandstone (sarsen)or flint-pebble conglomerate (puddingtone), widely distributed in southern England

solifluction: a gravity-driven, mass-wasting process due to repeated wetting-drying or freeze-thaw in periglacial regions *specific gravity*: the density (mass per unit volume) of a substance relative to water

squibs: small pieces of clay, moulded into a form such as a star, stuck using clay slip onto a pot as decoration, typically found on medieval wares

speleothems: laminated bodies of calcite (dripstone, stalactites, stalagmites) precipitated in caves from lime-rich waters

stope: a steeply inclined, slot-like working chamber within a mine

tidal range: the vertical separation at a coastal site between the levels of high tide and low tide, varying with geographical location, the rotation of the Earth and the phase of the moon. The range is normally greatest when the Sun, Earth and Moon are aligned and least when the Moon acts at right-angles to the Sun-Earth axis. The tidal range is subject to several periodicities, ranging from semidiurnal to 18.6 years. The range measures centimetres to decimetres in the open oceans but increases as seas become more confined

unit cell: the smallest geometrical arrangement of atoms that defines a mineral

varves: layers on a millimetre-centimetre scale of alternately fine and coarse, or organic-rich and organic-poor, sediment deposited in lakes and floodplains, of seasonal or sub-seasonal origin

whalebacks: ice-smoothed ribs of rock elongated parallel with the direction of ice- flow

zooplankton: various small invertebrates, including many larvae, that live in ocean surface waters and contribute to the base of the food chain

Index